Beast ACADEMY

By Art of Problem Solving

MATH
PRACTICE
2B

Jason Batterson

Kyle Guillet

Chris Page

Published by: AoPS Incorporated
                       10865 Rancho Bernardo Rd Ste 100
                       San Diego, CA 92127-2102
                       info@BeastAcademy.com

ISBN: 978-1-934124-33-8

Written by Jason Batterson, Kyle Guillet, and Chris Page
Book Design by Lisa T. Phan
Illustrations by Erich Owen
Grayscales by Greta Selman

Visit the Beast Academy website at BeastAcademy.com.
Visit the Art of Problem Solving website at artofproblemsolving.com.
Printed in the United States of America.
2019 Printing.

Contents:

This is Practice Book 2B in the Beast Academy level 2 series.

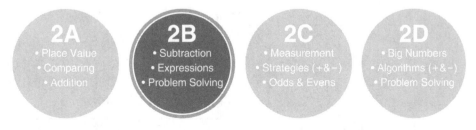

2A
• Place Value
• Comparing
• Addition

2B
• Subtraction
• Expressions
• Problem Solving

2C
• Measurement
• Strategies (+&−)
• Odds & Evens

2D
• Big Numbers
• Algorithms (+&−)
• Problem Solving

For more resources and information, visit BeastAcademy.com.

This is Beast Academy Practice Book 2B.

Each chapter of this Practice book corresponds to a chapter from Beast Academy Guide 2B.

MATH PRACTICE 2B

MATH GUIDE 2B

The first page of each chapter includes a recommended sequence for the Guide and Practice books.

You may also read the entire chapter in the Guide before beginning the Practice chapter.

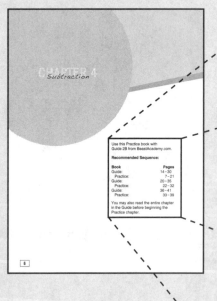

CHAPTER 4
Subtraction

Use this Practice book with Guide 2B from BeastAcademy.com.

Recommended Sequence:

Book	Pages
Guide:	14 – 30
Practice:	7 – 21
Guide:	31 – 35
Practice:	22 – 32
Guide:	36 – 41
Practice:	33 – 39

You may also read the entire chapter in the Guide before beginning the Practice chapter.

Some problems in this book are very challenging. These problems are marked with a ★. The hardest problems have two stars!

Every problem marked with a ★ has a *hint!*

Hints for the starred problems begin on page 108.

54.
★

42   Guide Pages: 39-43

Some pages direct you to related pages from the Guide.

None of the problems in this book require the use of a calculator.

Solutions are in the back, starting on page 112.

A complete explanation is given for every problem!

CHAPTER 4
Subtraction

Use this Practice book with
Guide 2B from BeastAcademy.com.

Recommended Sequence:

Book	Pages:
Guide:	14-30
Practice:	7-21
Guide:	31-35
Practice:	22-32
Guide:	36-41
Practice:	33-39

You may also read the entire
chapter in the Guide before
beginning the Practice chapter.

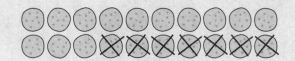

EXAMPLE | There are 20 cookies in a jar. After 7 of the cookies are eaten, how many cookies are left?

$$20 - 7 = 13$$

We use subtraction to take away.

Taking 7 away from 20 leaves 13. So, there are *13* cookies left in the jar.

PRACTICE | Fill the blanks to answer each subtraction problem below.

1. There are 12 frogs on a lily pad. After 3 of the frogs hop into the water, how many frogs are left on the lily pad?

    ___ − ___ = ___

2. A pitcher holds 15 cups of water. Priscilla pours 8 cups of water from the pitcher into glasses. How many cups of water are left in the pitcher?

    ___ − ___ = ___

3. Last week, Tyrone lost 9 of his 17 toy trucks in the sandbox. How many toy trucks does Tyrone have left?

    ___ − ___ = ___

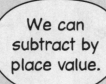

We can subtract by place value.

EXAMPLE | What is 65−23?

$$65 - 23 = 42$$

6 tens minus 2 tens is 4 tens.

5 ones minus 3 ones is 2 ones.

4 tens and 2 ones is **42**.

PRACTICE | Fill the blanks to solve each subtraction problem below.

4. 74 − 21 = $\underset{\text{tens}}{7}$ $\underset{\text{ones}}{4}$ − $\underset{\text{tens}}{2}$ $\underset{\text{ones}}{1}$ = $\underset{\text{tens}}{\quad}$ $\underset{\text{ones}}{\quad}$

5. 69 − 34 = $\underset{\text{tens}}{6}$ $\underset{\text{ones}}{9}$ − $\underset{\text{tens}}{3}$ $\underset{\text{ones}}{4}$ = $\underset{\text{tens}}{\quad}$ $\underset{\text{ones}}{\quad}$

6. 25 − 13 = _____

7. 79 − 36 = _____

8. 265 − 51 = _____

9. 280 − 120 = _____

10. 765 − 362 = _____

11. 957 − 926 = _____

PRACTICE | Fill the blanks in each subtraction problem below.

12. $46 - \underline{\hspace{1cm}} = 35$ **13.** $97 - \underline{\hspace{1cm}} = 63$ **14.** $746 - \underline{\hspace{1cm}} = 243$

15. $\underline{\hspace{1cm}} - 24 = 13$ **16.** $\underline{\hspace{1cm}} - 42 = 46$ **17.** $\underline{\hspace{1cm}} - 315 = 210$

18. Ms. Shloop has 87 hard candies. She gives 1 candy to each   **18.** _____
of her 36 students. How many candies does she have left?

19. Ralph used 232 stickers for an art project. 110 of the   **19.** _____
★ stickers were from a small pad. The rest were from a big
pad that started with 675 stickers. How many stickers are
left on the big pad?

20. Cammie and Nellie each write a 3-digit number.   **20.** _____
★ Cammie's hundreds digit is 3 more than Nellie's hundreds digit.
Cammie's tens digit is 2 more than Nellie's tens digit.
Cammie's ones digit is 1 more than Nellie's ones digit.
What is Cammie's number minus Nellie's number?

SUBTRACTION

Breaking

EXAMPLE | What is 42−17?

We can't take 7 ones away from 2 ones.

$$42 - 17 = \ ?$$

But, we can break one of the tens in 42 to make 10 ones.
4 tens and 2 ones is the same as 3 tens and 12 ones.

$$\overset{3\ \ 12}{\cancel{42}} - 17 = 25$$

3 tens minus 1 ten is 2 tens.
12 ones minus 7 ones is 5 ones.
2 tens and 5 ones is 25. So, 42−17 = **25**.

PRACTICE | Fill the blanks to solve each subtraction problem below.

21. 33−15 = $\underset{tens}{3}\ \underset{ones}{3}$ − $\underset{tens}{1}\ \underset{ones}{5}$

= $\underset{tens}{2}\ \underset{ones}{13}$ − $\underset{tens}{1}\ \underset{ones}{5}$

= $\boxed{18}$

22. 41−26 = $\underset{tens}{4}\ \underset{ones}{1}$ − $\underset{tens}{2}\ \underset{ones}{6}$

= $\underset{tens}{3}\ \underset{ones}{11}$ − $\underset{tens}{2}\ \underset{ones}{6}$

= $\boxed{15}$

23. 86−37 = $\underset{tens}{8}\ \underset{ones}{6}$ − $\underset{tens}{3}\ \underset{ones}{7}$

= $\underset{tens}{7}\ \underset{ones}{16}$ − $\underset{tens}{3}\ \underset{ones}{7}$

= $\boxed{49}$

24. 72−48 = $\underset{tens}{7}\ \underset{ones}{2}$ − $\underset{tens}{4}\ \underset{ones}{8}$

= $\underset{tens}{6}\ \underset{ones}{12}$ − $\underset{tens}{4}\ \underset{ones}{8}$

= $\boxed{64}$

PRACTICE | Solve each problem below.

25. $51 - 13 =$ ___38___

26. $82 - 25 =$ ___57___

27. $94 - 46 =$ ___48___

28. $63 - 38 =$ ___25___

29. ★ $235 - 73 =$ ___162___

30. ★ $317 - 180 =$ ___137___

31. Polly Pandakeet has 54 crackers. He eats 26 of them for lunch. How many crackers does Polly Pandakeet have left?

31. ___28___

32. Ms. Melody teaches 215 students in her orchestra classes. Yesterday, 72 of her students left school to play in a concert. How many of Ms. Melody's students **did not** play in the concert?

32. ___143___

33. ★ Lizzie and Alex each write a two-digit number. Lizzie's tens digit is 2 **more** than Alex's tens digit. Lizzie's ones digit is 2 **less** than Alex's ones digit. What is Lizzie's number minus Alex's number?

33. ___18___

$42 - 24$

EXAMPLE | In a class of 30 monsters, 18 have horns. How many monsters in the class do not have horns?

The 18 monsters with horns plus the number of monsters without horns gives 30 total monsters.

$$18 + \boxed{} = 30$$

Since $18 + \underline{12} = 30$, there are **12** monsters who do not have horns.

– or –

If we take away the 18 monsters with horns, only monsters without horns will be left.

$$30 - 18 = \boxed{}$$

Since $30 - 18 = \underline{12}$, there are **12** monsters who do not have horns.

We can use subtraction to find the missing number in a sum!

PRACTICE | Answer each question below.

34. Stu flips 21 coins. Each coin lands either heads or tails. If 13 coins landed heads, how many coins landed tails?

34. _____

35. Mary spent 55 minutes walking and jogging. If she spent 30 minutes jogging, how many minutes did she spend walking?

35. _____

36. Together, Brett and Chuck have 83 dollars. Brett has only two 20-dollar bills. How many dollars does Chuck have?

36. _____

37. A 32-foot rope is cut into 2 pieces. One piece is 14 feet long. How many feet long is the other piece?

37. _____

EXAMPLE | Fill in the missing numbers in the boxes below.

26 + 65 = 91   65 + 26 = ☐

91 − 26 = ☐   91 − 65 = ☐

Addition and subtraction are related.

Knowing a sum can help you subtract.

Since 26 plus 65 is 91, we know 65 plus 26 is also 91.
So, if we take away 26 from 91, we are left with 65,
and if we take away 65 from 91, we are left with 26.

26 + 65 = 91   65 + 26 = **91**

91 − 26 = **65**   91 − 65 = **26**

PRACTICE | Fill the empty boxes in each problem below so that each uses *the same three numbers* to make *four different statements.*

38. 27 + 56 = ☐

56 + ☐ = 83

☐ − 56 = 27

83 − ☐ = 56

39. 88 + 44 = ☐

44 + ☐ = 132

132 − ☐ = 88

☐ − 88 = ☐

40. ☐ − 58 = 123

☐ − 123 = 58

123 + 58 = ☐

58 + ☐ = ☐

41. ★ 39 + ☐ = ☐

☐ + ☐ = 64

☐ − 25 = ☐

☐ − ☐ = ☐

In a **Three-Four** puzzle, *three* different numbers are used to make *four* related addition and subtraction statements.

EXAMPLE | In the Three-Four puzzle below, fill the empty squares so that only three different numbers make all four statements.

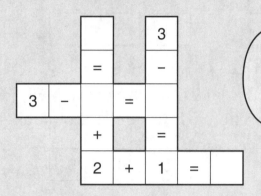

We read the statements from left to right and top to bottom.

Since 1+2 is 3, we can also say that 3 is 1+2. 3=1+2 means the same thing as 1+2=3.

We fill the empty squares as shown below.

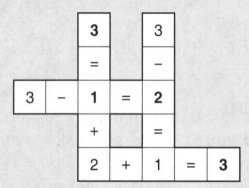

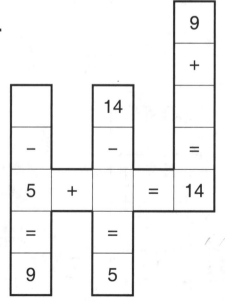

PRACTICE | In each Three-Four puzzle, fill the empty squares so that only three different numbers make all four statements.

42.

```
7
–
4  +  3  =  
=     +     –
            3
      =     =
      7
```

43.

```
            9
            +
      14
–     –     =
5  +     =  14
=     =
9     5
```

44.

28	−	12	=	16
−				
16	+	12	=	28
=				
12	+	16	=	28

45.

24		87		
+		−		
63	+	24	=	87
=	▓	=		
87	−	63	=	24

46.

56				
+				
18		74		
=		−		
74	=	18	+	56
		=		
74	−	56	=	18

47.

222				
+				
111		111		
=		+		
333	−	222	=	111
		=		
333	−	111	=	222

48.

562		237		
−		+		
237	562	−	325	=	237
=		=		
325	+	237	=	562

49.

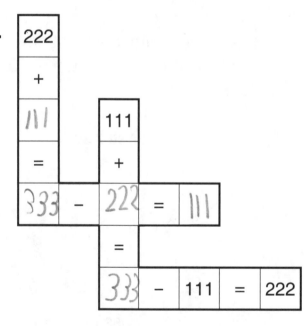

238	+	686	=	924
+	▓	−		
238	▓	686		
=	▓	=		
924	−	238	=	686

In an **X-Out** puzzle, we cross out numbers so that the sum of the remaining numbers in each row and column equals the target sum.

EXAMPLE | Cross out numbers in the grid to the right so that every row and column has a sum of 30.

Sum: 30

6	12	13	17
4	18	8	5
15	19	9	6
11	12	10	7

In the top row, 13 and 17 are the only numbers we can add to get a sum of 30. So, we cross out 6 and 12. We also circle 13 and 17 to remind ourselves that these numbers must be used.

Sum: 30

~~6~~	~~12~~	(13)	(17)
4	18	8	5
15	19	9	6
11	12	10	7

In the right column, we know we must use 17, so the remaining numbers must have a sum of 30 − 17 = 13. Only 6 and 7 have a sum of 13. So, we cross out 5, and circle 6 and 7.

Sum: 30

~~6~~	~~12~~	(13)	(17)
4	18	8	~~5~~
15	19	9	(6)
11	12	10	(7)

We use similar strategies to finish the puzzle, as shown below.

Sum: 30

~~6~~	~~12~~	(13)	(17)
(4)	(18)	(8)	~~5~~
(15)	~~19~~	(9)	(6)
(11)	(12)	~~10~~	(7)

PRACTICE | Cross out numbers in each grid below so that every row and column has the target sum.

50. Sum: 10

~~2~~	~~2~~	(5)	(5)
(2)	(3)	~~3~~	(5)
(4)	(3)	(3)	~~2~~
(4)	(4)	(2)	~~2~~

51. Sum: 22

(15)	(4)	(3)	~~9~~
~~5~~	~~11~~	(4)	(18)
~~3~~	(12)	(10)	~~1~~
(7)	(6)	(5)	(4)

52. Sum: 88

(22)	14	22	44
(22)	32	34	44
(22)	34	32	44
(22)	22	64	44

PRACTICE | Cross out numbers in each grid below so that every row and column has the target sum.

53. **Sum**: 112

4	106	2	103
101	6	102	5
7	108	1	104
107	8	109	3

54. **Sum**: 123

100	20	100	3
3	20	20	100
20	3	100	3
3	100	3	20

55. **Sum**: 8

4	1	2	3
4	2	4	1
4	4	4	4
2	3	4	1

56. **Sum**: 99

50	49	20	20	50
10	10	20	20	39
10	10	39	30	30
39	40	40	30	30
50	40	40	19	50

57. ★ **Sum**: 28

10	9	9	9	19
6	26	11	17	9
6	8	23	5	9
6	20	2	14	9
16	6	6	6	10

58. ★★ **Sum**: 8

4	4	8	8	8
2	4	4	4	8
2	2	4	4	8
2	2	2	4	4
1	2	2	2	4

59. ★★ **Sum**: 25

10	10	5	5	5
5	5	5	10	5
5	10	10	10	5
5	5	10	10	5
10	10	10	5	5

Print more of these puzzles at BeastAcademy.com.

EXAMPLE | How much greater is 84 than 31?

We are looking for the number that can be added to 31 to give us 84.

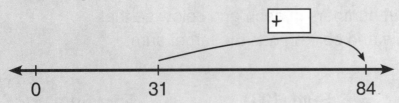

We can find the missing number in this addition problem using subtraction: $84 - 31 = \boxed{53}$.

So, 84 is greater than 31 by **53**.

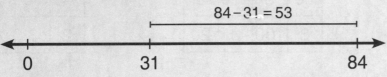

*We can use subtraction to find the difference between two numbers. So, the result of subtraction is called a **difference**.*

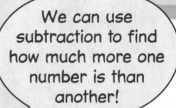

We can use subtraction to find how much more one number is than another!

In this book, the **difference** is always the bigger number minus the smaller number.

PRACTICE | Answer each question below.

60. How much greater is 65 than 23?

60. _____

61. What is the difference between 555 and 234?

61. _____

62. Circle the two numbers below that have the greatest difference.

64   19   108   31   156   88

63. What is the greatest possible difference between a 3-digit number and a 2-digit number?

63. _____

EXAMPLE

Urg has 148 books.
Yarg has 113 books.
How many more books does
Urg have than Yarg?

We want to know how much more 148 is than 113. In other words, we want to find the difference between 148 and 113.

To find a difference, we subtract. So, Urg has 148 − 113 = **35** more books than Yarg.

PRACTICE | Answer each question below.

64. Jack fetches 26 pails of water. Jill fetches 43 pails of water. How many more pails did Jill fetch than Jack?

64. _____

65. Knobby Knoll is 350 feet tall. Humpback Hill is 460 feet tall. How many feet taller is Humpback Hill than Knobby Knoll?

65. _____

66. Rebecca is 9 years older than Shelly, who is 7 years older than Tamika. What is the difference in years between Rebecca's age and Tamika's age?

66. _____

67. ★ Jeremy reads a 222-page book, then a 328-page book. Richard reads a 236-page book, then a 348-page book. How many more pages did Richard read than Jeremy?

67. _____

EXAMPLE | What is 132−95?

Instead of subtracting by place value, we can find the difference between 132 and 95 on the number line.

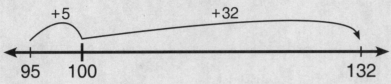

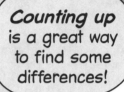

Counting up is a great way to find some differences!

100 is 5 more than 95.
132 is 32 more than 100.
So, 132 is 5+32 = 37 more than 95.

This means that 132−95 = **37**.

PRACTICE | Use the number lines below to find each difference.

68.

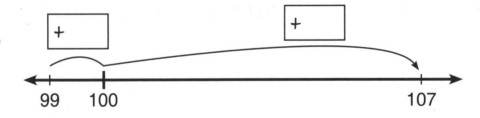

107−99 = _____

69.

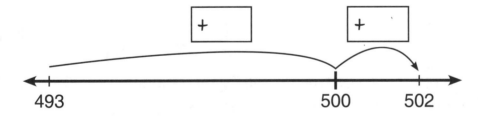

502−493 = _____

70.

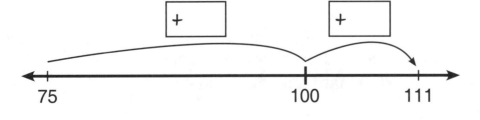

111−75 = _____

PRACTICE | Use the number lines below to find each difference.

71.

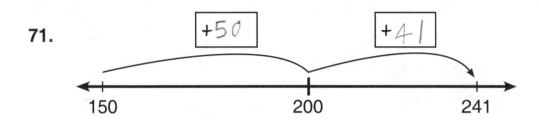

$241 - 150 =$ ___11___

72.

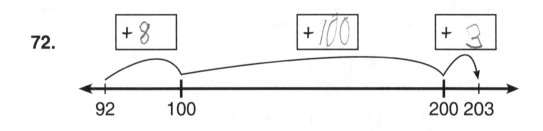

$203 - 92 =$ _____

73.

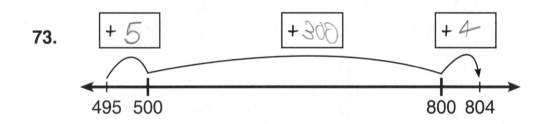

$804 - 495 =$ _____

PRACTICE | Find each difference below.

74. $125 - 97 =$ _____

75. $240 - 185 =$ _____

76. $700 - 290 =$ ___410___

77. $231 - 175 =$ _____

78. $543 - 350 =$ ___193___

79. $909 - 591 =$ _____

In a **Hop-Along** puzzle, we use distance clues to draw the path that Hoppy the Humblebug takes to visit the points shown on the number line once each.

EXAMPLE | Draw the hops taken by Hoppy on the number line below to travel the given distances, in order.

Distances: 16, 47

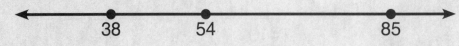

38   54   85

The first hop is a distance of 16. Only 54 and 38 are 16 apart: $54 - 38 = 16$.

So, Hoppy either starts at 38 and hops to 54, or starts at 54 and hops to 38. We draw this hop on the number line as shown.

Distances: 16, 47

38   54   85

The next distance is 47. Only 85 and 38 are 47 apart: $85 - 38 = 47$. So, Hoppy must start at 54, hop to 38, then hop to 85.

We complete Hoppy's path as shown. We draw a circle at the start of his trip, and an arrow at the end of his trip.

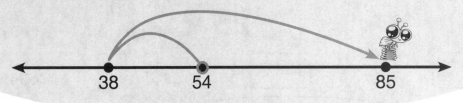

38   54   85

PRACTICE | On each number line, draw a path of hops that travels the given distances, in order. Draw a circle at each path's start, and an arrow at each path's end.

80. **Distances**: 41, 25

29   45   70

81. **Distances**: 26, 53

26   53   79

82. **Distances**: 43, 79, 60

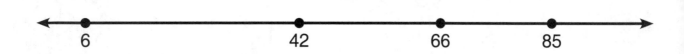

83. **Distances**: 10, 71, 81

84. **Distances**: 58, 11, 41, 8

85. ★ **Distances**: 35, 37, 33, 39

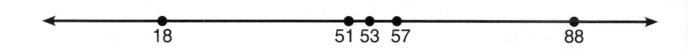

86. ★ **Distances**: 39, 40, 39, 12

EXAMPLE | Alex has 250 sheets of paper. He gives Grogg 99 sheets. How many sheets of paper does Alex have left?

Sometimes it's easier to take away a little extra, then add it back!

To make the subtraction easier, Alex can give Grogg 100 sheets, then take 1 back. This is the same as giving Grogg 99 sheets of paper. So,

$$250-99=250-100+1.$$

$250-100=150$, and $150+1=151$. So, Alex has **151** sheets left.

PRACTICE | Answer each question below.

87. Brian has 350 dollars. Brian gives Ted a 100-dollar bill, then takes back a 5-dollar bill from Ted. Circle **both** of the values below that describe how many dollars Brian has now.

$350-95$ $350-100-5$ $350-100+5$ $350-105$

88. There are 284 students in a gym. 100 students leave to go to lunch. Then, 12 of those students return to the gym. Circle **both** of the values below that give the number of students who are still in the gym.

$284-112$ $284-88$ $284-100-12$ $284-100+12$

89. Circle the value below that is equal to $755-198$.

$755+200+2$ $755-200+2$ $755-200-2$ $755+200-2$

PRACTICE | Fill in the blanks below.

90. Subtracting 96 is the same as subtracting 100, then adding _____.

91. Subtracting 190 is the same as subtracting _____, then adding 10.

92. Subtracting _____ is the same as subtracting 50, then adding 2.

93. $261 - 94 = $ | 261 | $-$ | 100 | $+$ | ___ | $=$ | ___ |

94. $72 - 39 = $ | 72 | $-$ | ___ | $+$ | 1 | $=$ | ___ |

95. $450 - 195 = $ | 450 | $-$ | ___ | $+$ | ___ | $=$ | ___ |

96. $123 - 97 = $ _____   **97.** $383 - 49 = $ _____   **98.** $650 - 195 = $ _____

99. $333 - 75 = $ _____   **100.** $992 - 299 = $ _____   **101.** $868 - 390 = $ _____

So far, we've learned to subtract by place value...

...by counting up...

...or by taking away a little extra, then giving it back!

Which of these do you use most?

PRACTICE | Find each difference below.

102. 355 − 96 = _____

103. 355 − 122 = _____

104. 355 − 275 = _____

105. 517 − 486 = _____

106. 517 − 198 = _____

107. 517 − 214 = _____

108. 876 − 765 = _____

109. 876 − 780 = _____

110. 876 − 90 = _____

111. 539 − 470 = _____

112. 539 − 295 = _____

113. 539 − 226 = _____

114. 707 − 604 = _____

115. 707 − 659 = _____

116. 707 − 280 = _____

Fill each blank below with a digit.

For example, $\boxed{5}8$ is the number 58.

PRACTICE | Fill the empty blanks below to make each statement true.

117. $19 - \boxed{} = 12$

118. $25 - 1\boxed{} = 12$

119. $47 - \boxed{}3 = 14$

120. $\boxed{}7 - 3\boxed{} = 55$

121. $9\boxed{} - \boxed{}6 = 12$

122. $8\boxed{} - 34 = \boxed{}5$

123. $34 - \boxed{} = \boxed{}8$

124. $\boxed{}1 - 8 = 4\boxed{}$

125. ★ $74 - \boxed{}5 = 1\boxed{}$

126. ★ $6\boxed{} - \boxed{}7 = 34$

In a **Subtractile** puzzle, each △ is the difference of the two ☐'s it touches.

Subtraction Subtractiles

EXAMPLE Fill each empty shape with a whole number to complete the Subtractile puzzle to the right.

First, we fill the two empty triangles.
46 − 25 = 21 and 46 − 19 = 27. So, we fill
the two empty triangles with 21 and 27.

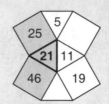 

The difference of the empty square and 25 is 5. This
means the empty square is 5 more or 5 less than 25.
So, it is 25 + 5 = 30 or 25 − 5 = 20.

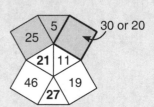

The difference of the empty square and 19 is 11. This
means the empty square is 11 more or 11 less than 19.
So, it is 19 + 11 = 30 or 19 − 11 = 8.

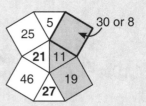

30 is the only number that correctly gives both
differences. So, the empty square is 30.

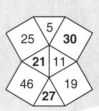

PRACTICE Fill each empty shape with a whole number
to complete the Subtractile puzzles below.

127.

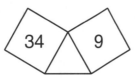

128.

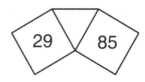

129.

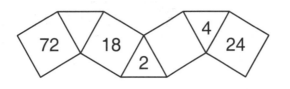

130.

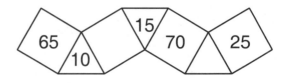

PRACTICE | Fill each empty shape with a whole number to complete the Subtractile puzzles below.

131.

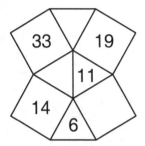

132.

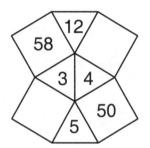

133.
★

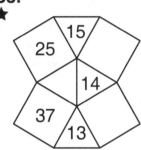

134.

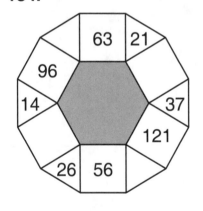

135.
★

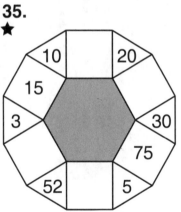

136.
★
★

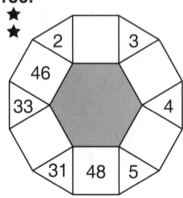

137.
★
★

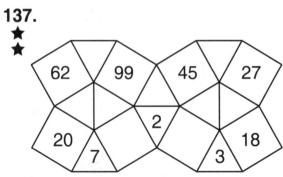

138.
★
★

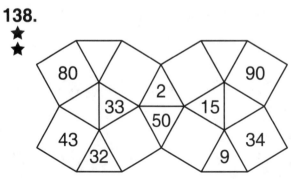

SUBTRACTION Changing a Difference

EXAMPLE | 324−87 = 237. What is 334−87?

We can use 324−87 to help us find 334−87.

In 334−87, we start with 10 more than in 324−87, but we subtract the same amount.

Starting with 10 more makes our result greater by 10.

324−87 = 237. So, we have
334−87 = 237+10 = **247**.

– or –

We consider the difference on the number line.

334 is 10 units farther from 87 than 324 is from 87. So, 334−87 is 10 more than 324−87.

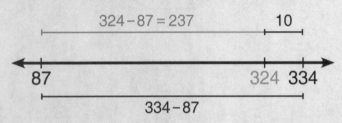

Since 324−87 = 237, we have 334−87 = 237+10 = **247**.

How does changing the numbers in a subtraction problem change the difference?

PRACTICE | Answer each question below.

139. 613−548 = 65. What is 614−548?

139. _____

140. 921−267 = 654. What is 921−367?

140. _____

141. 427−168 = 259. What is 427−148?

141. _____

142. ★ 723−237 = 486. What is 823−137?

142. _____

143. ★ 632−466 = 166. What is 622−476?

143. _____

PRACTICE | Answer each question below.

144. How much *greater* is 538 − 159 than 538 − 179?   **144.** _____

145. How much *less* is 319 − 144 than 719 − 144?   **145.** _____

146. Greg is exactly 13 years older than Meg. In 8 years, how many years older will Greg be than Meg?   **146.** _____

147. The difference between two numbers is 50. If you add 6 to the big number and subtract 6 from the small number, what is the difference between the two new numbers?   **147.** _____

148. Polly has 10 more grapes than Molly. If Polly gives 3 grapes ★ to Molly, how many more grapes will Polly have than Molly?   **148.** _____

149. How many pairs of *two-digit* numbers have a difference of 80?   **149.** _____
★

EXAMPLE | What is 115−96?

We can add or subtract the **same amount** from two numbers without changing their difference.

Adding 4 to both 115 and 96 makes the subtraction easier.

We have 115−96 = 119−100 = **19**.

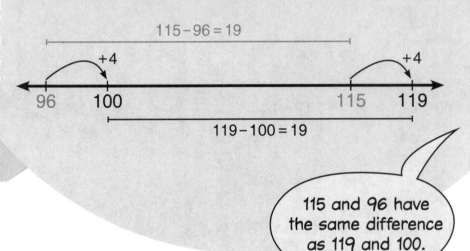

Adding or subtracting the **same amount** from both numbers in a subtraction problem doesn't change the difference!

115 and 96 have the same difference as 119 and 100.

PRACTICE | Fill the blanks to solve each problem below.

150. 85−38 has the same difference as _____ −40.

151. 85−38 = _____

152. 271−94 has the same difference as _____ −100.

153. 271−94 = _____

154. 183−57 has the same difference as _____ −60.

155. 183−57 = _____

156. ★ Circle *every* pair of numbers below that has the same difference as 241−82.

240−81   250−91   259−100   239−84   200−41

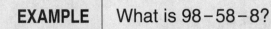

We can **add** numbers in any order we want...

...but when we **subtract**, we have to be careful!

EXAMPLE | What is 98−58−8?

When subtracting more than one number, working from **left to right** always gives the correct result.

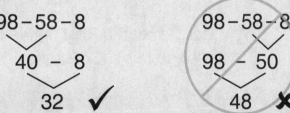

$$98-58-8$$
$$40 \quad - \quad 8$$
$$32 \quad \checkmark$$

$$98-58-8$$
$$98 \quad - \quad 50$$
$$48 \quad \times$$

So, 98−58−8 = **32**.

PRACTICE | Solve each problem below.

157. 32 − 12 − 2 = _____

158. 148 − 116 − 16 = _____

159. 297 − 143 − 43 = _____

160. 201 − 99 − 40 = _____

161. 188 − 44 − 33 − 11 = _____

162. 285 − 25 − 25 − 25 = _____

163. 168 − 68 − 70 − 29 = _____

164. 345 − 45 − 67 − 89 = _____

Subtracting left-to-right always works, but sometimes we can be more clever!

EXAMPLE Marc begins a 320-page book. He reads 44 pages in the morning, then 56 pages at night. How many pages does he have left to read?

After Marc reads 44 pages, he will have $320-44=276$ pages left to read. Then, after reading 56 more pages, he will have $276-56=\textbf{220}$ pages left.

– or –

Marc has read a total of $44+56=100$ pages from his 320-page book. So, he has $320-100=\textbf{220}$ pages left.

Since $44+56=100$, taking away 44 then taking away 56 is the same as taking away 100 all at once!

So, $320-44-56=320-100$.

PRACTICE Solve each problem and fill in the blanks below.

165. Alisha has 189 dollars. She buys a 27-dollar pair of headphones, then buys a 23-dollar pair of sunglasses. How many dollars does Alisha have left?

165. _____

166. *Gone with the Beasts* is a 262-minute-long movie. Snorg watches 75 minutes before pausing for lunch, then watches another 25 minutes. How many minutes are left in the movie?

166. _____

167. $84-38-12=\boxed{84}-\boxed{}$

168. $84-38-12=$ _____

169. $511-22-33-44=\boxed{511}-\boxed{}$

170. $511-22-33-44=$ _____

Sometimes, it's easier to subtract a number in parts.

EXAMPLE | Mary has read 89 pages of a 359-page book. How many pages does she have left to read?

Reading 89 pages is the same as reading 59 pages, then 30 more pages. So, to subtract 89, we can subtract 59, then subtract 30 more.

$$359 - 89 = 359 - 59 - 30.$$

$359 - 59 = 300$, and $300 - 30 = 270$.
So, Mary has **270** pages left to read.

PRACTICE | Solve each problem and fill in the blanks below.

171. There are 137 jelly beans in a bag. Alfred takes out 45 beans. Of these 45 beans, he gives 37 to Sam, then eats the other 8. How many jelly beans are left in the bag?

171. _____

172. 352 birds sit in a tree. 167 of them fly away. If 152 of the birds who fly away are black, and 15 are blue, how many birds are still in the tree?

172. _____

173. Circle *every* value below that is equal to $755 - 88$.

$$755 - 55 - 33 \qquad 755 - 55 + 33 \qquad 755 - 5 - 83 \qquad 755 - 44 + 44$$

174. $174 - 86 = \boxed{174} - \boxed{74} - \boxed{}$

175. $174 - 86 = $ _____

176. $311 - 66 = \boxed{311} - \boxed{11} - \boxed{}$

177. $311 - 66 = $ _____

EXAMPLE | Ida has a basket of 122 eggs. She gives 56 eggs to her dad, then gives 22 eggs to her mom. How many eggs are left in Ida's basket?

It doesn't matter if Ida gives eggs to her dad and then to her mom, or if she gives eggs to her mom and then to her dad.

No matter what order Ida gives away her eggs, she will be left with the same number of eggs in her basket at the end.

So, 122−56−22 is equal to 122−22−56. Since subtracting 22 first is easier, we compute 122−22−56.

$$122-22-56$$
$$100 \; - \; 56$$
$$44$$

So, **44** eggs are left in Ida's basket.

> *Careful when changing the order of subtraction!*
>
> *We can't reorder the number we're subtractin' from.*
>
> 122−56−22 is **not** equal to 56−122−22.

PRACTICE | Answer each question below.

178. There are 182 apples in a tree. Eve picks 35 apples in the morning, then 82 apples in the evening. How many apples are still on the tree?

178. _____

179. Juan spent 45 dollars, then 17 dollars, then 35 dollars at three different stores. He started with 117 dollars. How many dollars does he have left after visiting all three stores?

179. _____

180. $71-45-11 = $ _____

181. $465-9-365 = $ _____

182. $234-150-34 = $ _____

183. $225-37-75 = $ _____

The problems on these next three pages are tough!

PRACTICE | Answer each question below.

184. Subtract: $128-64-32-16-8-4-2-1 =$ _____.

185. Removing 654 pounds of sand from a truck leaves 321 pounds of sand in the truck. How many pounds of sand were in the truck before any sand was removed?

185. _____

186. Genie subtracts the four largest two-digit numbers from 400. What result does she get?

186. _____

187. ★ Write the **same** number in each blank below to make a true statement.

$$40-\text{\_\_\_\_}-\text{\_\_\_\_}-\text{\_\_\_\_}=4$$

PRACTICE | Answer each question below.

188. Emily is counting by 7's: 7, 14, 21, 28, 35, and so on. What is   **188.** _____
★   the difference between the 30th number she says and the 33rd
number she says?

189. In the subtraction problem below, both blanks are missing the *same digit*.
★   Fill each blank with the missing digit.

$$84 - 2\square = 5\square$$

190. If each blank below is filled with a *different* three-digit   **190.** _____
★   number, what is the largest result that is possible?

$$\underline{\qquad} - \underline{\qquad} - \underline{\qquad}$$

191. In the statement below, each shape stands for a *different* number.
★
★

$$\square - \triangle = \bigcirc$$

If the above statement is true, circle every statement below that is also true.

$$\square - \bigcirc = \triangle \qquad\qquad \triangle + \square = \bigcirc$$

$$\bigcirc + \triangle = \square \qquad\qquad \bigcirc - \triangle = \square$$

$$\triangle - \bigcirc = \square \qquad\qquad \triangle + \bigcirc = \square$$

PRACTICE | Answer each question below.

192. ★ What is the smallest number that can be subtracted from 456 to get a two-digit result?

192. _____

193. ★ When Stuart stands on a 19-inch stool, he is as tall as his dad, who is 74 inches tall. How many inches taller is Stuart than the stool?

193. _____

194. ★ Winnie lists all of the numbers that use the digits 1, 2, and 3 once each. What is the difference between the largest and the smallest number in Winnie's list?

194. _____

195. ★ Hans has a calculator with only the six buttons shown. What is the smallest number of button presses Hans needs to get his calculator to read 91?

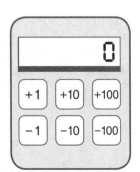

195. _____

CHAPTER 5
Expressions

Use this Practice book with
Guide 2B from BeastAcademy.com.

Recommended Sequence:

Book	Pages:
Guide:	44-54
Practice:	41-55
Guide:	55-59
Practice:	56-61
Guide:	60-73
Practice:	62-71

You may also read the entire
chapter in the Guide before
beginning the Practice chapter.

An **expression** uses numbers and operations like + and - to stand for a value.

EXAMPLE | Evaluate $15-6+8-5-1$.

$$15-6+8-5-1$$
$$= 9+8-5-1$$
$$= 17-5-1$$
$$= 12-1$$
$$= \mathbf{11}.$$

When we **evaluate** an expression, we find its value.

When adding and subtracting, we always get the correct value by working from left to right.

PRACTICE | Evaluate each expression below.

1. $7+3-5 =$ _____

2. $19-8+11 =$ _____

3. $34-13+9 =$ _____

4. $275-204+111 =$ _____

5. $70+50-12 =$ _____

6. $200-79+55 =$ _____

7. $19-16+12-14 =$ _____

8. $59+30-17+13 =$ _____

9. $580-30-140+60 =$ _____

10. $203+177-70-70 =$ _____

Fill the empty boxes so that the statements from left-to-right and from top-to-bottom are all correct.

EXAMPLE | Complete the Cross-Number puzzle below.

20	–	4	–	3	=	
+	■	–	■	+	■	+
15	+	2	–	8	=	
–	■	–	■	+	■	–
4	–	1	+	7	=	
=	■	=	■	=	■	=
	–		–		=	

Across:

$20-4-3=\boxed{13}$.

$15+2-8=\boxed{9}$.

$4-1+7=\boxed{10}$.

$\boxed{31}-\boxed{1}-\boxed{18}=\boxed{12}$.

Down:

$20+15-4=\boxed{31}$.

$4-2-1=\boxed{1}$.

$3+8+7=\boxed{18}$.

$\boxed{13}+\boxed{9}-\boxed{10}=\boxed{12}$.

20	–	4	–	3	=	**13**
+	■	–	■	+	■	+
15	+	2	–	8	=	**9**
–	■	–	■	+	■	–
4	–	1	+	7	=	**10**
=	■	=	■	=	■	=
31	–	**1**	–	**18**	=	**12**

PRACTICE | Complete each Cross-Number puzzle below.

11.

1	+	2	+	3	=	
+	■	+	■	+	■	+
8	–	3	–	3	=	
+	■	–		+	■	–
5	–	2	+	3	=	
=	■	=	■	=	■	=
	–		–		=	

12.

20	+	12	–	5	=	
–	■	–	■	+	■	–
6	+	3	+	5	=	
+	■	–	■	–	■	+
8	–	4	+	2	=	
=	■	=	■	=	■	=
	+		–		=	

PRACTICE | Complete each Cross-Number puzzle below.

13.

45	−	34	+	14	=	
−		−		+		−
15	+	25	−	19	=	
−		+		−		+
6	+	10	−	5	=	
=		=		=		=
	+		−		=	

14.

15	+	120	+	70	=	
+		−		−		−
45	+	98	+	17	=	
+		+		−		+
39	−	11	+	23	=	
=		=		=		=
	−		+		=	

15. ★

10	−	3	+		=	9
−		+		+		−
	+	7	−	6	=	5
+		+		−		+
12	−	8	−		=	2
=		=		=		=
	−		+	6	=	

16. ★

11	+		+	6	=	39
−		−		+		−
8	+		−		=	1
+		−		−		+
15	−	1	−		=	9
=		=		=		=
	+	12	+		=	

We can turn sentences into expressions.

PRACTICE | Solve each problem below.

17. Circle the expression below that stands for "the sum of 3, 4, and 5."

3+4+5   3−4+5   3+4−5   3−4−5

18. Herman takes 8 cookies from a jar, eats 3, then gives 4 to Edna.
Circle the expression that describes how many cookies Herman has now.

8+3+4   8−3+4   8−3−4   8+3−4

19. Lynn has 45 golf balls in a basket. She hits 35 of the balls into a field, then collects 33 balls and puts them back into the basket. Circle the expression that describes how many balls are now in the basket.

45+35+33   45+35−33   45−35+33   45−35−33

20. Circle the expression below that stands for "3 less than the sum of 4 and 5."

4+5+3   3−4+5   4+5−3   5−4−3

PRACTICE | Write an expression for each word problem below. Then, evaluate the expression.

21. Winnie builds a block tower 17 blocks high. She places 10 blocks on the top of the tower, then 7 blocks fall off. How many blocks tall is the tower now?

_____ = _____

22. Alex has 27 bow ties. He sells 15 of them, then buys 6 new ones. How many bow ties does Alex now have?

_____ = _____

23. 36 black birds and 43 red birds sit in a tree. 19 of the red birds fly away. Then, a flock of 30 blue birds lands in the tree. How many birds are in the tree now?

_____ = _____

24. What number is four more than the difference of nine and six?

_____ = _____

25. What number is three less than the sum of five and eight?

_____ = _____

In an **Expression Search**, we circle expressions that equal the target number when read from left-to-right or top-to-bottom.

Circled expressions cannot overlap. Every number must be part of one circled expression, but not every + and − symbol will be circled. Remember, a number by itself is also an expression!

Target: 6

```
10 −  4 +  5
 3 +  3 −  7 +  3
 +  6 −  2 −  1 +
11 −  5 +  1 +  6
 −  6 +  4 −  9 −
 8 −  1 −  1 −  3
   12 −  6 −  3
```

EXAMPLE | Complete the Expression Search puzzle on the right.

We circle the twelve expressions that equal 6 as shown.

```
(10 − 4) + (5
(3 + 3) − (7) + (3)
 + (6) − (2) − (1) +
(11 − 5) + (1) + 6
 − (6) + (4) − (9) −
(8 − 1 − 1) − (3)
  (12 − 6) − (3)
```

PRACTICE | Complete each Expression Search puzzle below.

26. Target: 9

```
    9 + 1 + 3
  8 + 1 + 7 + 4
  + 4 + 2 + 2 +
  2 + 9 + 2 + 2
  + 5 + 6 + 1 +
  7 + 5 + 4 + 3
    7 + 2 + 3
```

27. Target: 20

```
    6 + 36 − 16
 30 + 10 − 2 + 12
  − 7 − 20 − 15 −
  5 + 10 + 10 + 4
  − 7 + 20 − 5 +
  5 − 22 − 2 + 16
   36 − 20 + 4
```

46 Guide Pages: 44-49

Beast Academy Practice 2B

PRACTICE | Complete each Expression Search puzzle below.

28. Target: 14

```
   20 – 12 + 5
42 – 7 – 3 + 10
   – 6 + 6 + 9 –
28 – 7 + 2 + 5
   – 15 – 8 + 14 +
14 – 14 – 9 + 9
   1 + 13 + 1
```

29. Target: 6

```
   8 – 2 + 6
11 – 2 – 3 + 5
   – 6 + 12 + 9 +
4 + 2 – 5 – 4
   + 7 – 6 + 5 –
6 – 6 + 1 + 3
   1 + 6 – 2
```

30. Target: 100

```
   163 – 182 – 51
75 – 144 – 8 + 200
   + 52 – 56 + 49 –
25 – 44 – 92 + 80
   + 11 + 26 – 27 –
131 – 59 + 28 + 20
   63 + 37 + 73
```

31. Target: 121

```
   151 – 30 – 91
201 + 171 – 50 + 60
   – 11 + 110 + 80 +
10 + 80 – 91 – 61
   – 21 + 131 + 20 +
70 + 41 – 30 – 121
   100 – 10 – 30
```

32. ★ Target: 12

```
   8 + 4 + 8
18 – 4 – 2 + 10
   – 2 + 1 + 7 –
6 + 8 + 12 + 5
   + 6 + 11 – 5 +
6 + 8 – 2 – 7
   4 + 2 + 6
```

33. ★★ Target: 1

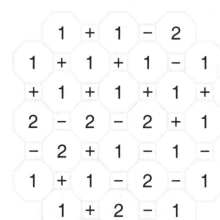

```
   1 + 1 – 2
1 + 1 + 1 – 1
   + 1 + 1 + 1 +
2 – 2 – 2 + 1
   – 2 + 1 – 1 –
1 + 1 – 2 – 1
   1 + 2 – 1
```

EXAMPLE | Fill each blank with + or − to make the statement below true.

$$27\ \boxed{}\ 9\ \boxed{}\ 3 = 33$$

There are four ways to fill the boxes.

$$27\ \boxed{+}\ 9\ \boxed{+}\ 3 = 39\ \text{✗}$$
$$27\ \boxed{+}\ 9\ \boxed{-}\ 3 = 33\ \text{✓}$$
$$27\ \boxed{-}\ 9\ \boxed{+}\ 3 = 21\ \text{✗}$$
$$27\ \boxed{-}\ 9\ \boxed{-}\ 3 = 15\ \text{✗}$$

Of these, only $27\ \boxed{+}\ 9\ \boxed{-}\ 3 = 33$ is true.

PRACTICE | Fill each blank with + or − to make the statements below true.

34. $23\ \boxed{}\ 2 = 21$

35. $15\ \boxed{}\ 6\ \boxed{}\ 7 = 28$

36. $30\ \boxed{}\ 9\ \boxed{}\ 2 = 19$

37. $25\ \boxed{}\ 25\ \boxed{}\ 50 = 50$

38. $90\ \boxed{}\ 36\ \boxed{}\ 34 = 92$

39. $90\ \boxed{}\ 36\ \boxed{}\ 34 = 88$

40. $20\ \boxed{}\ 6\ \boxed{}\ 3\ \boxed{}\ 8 = 21$

41. $300\ \boxed{}\ 30\ \boxed{}\ 30\ \boxed{}\ 40 = 200$

42. $18\ \boxed{}\ 16\ \boxed{}\ 14\ \boxed{}\ 12 = 8$

43. $38\ \boxed{}\ 28\ \boxed{}\ 25\ \boxed{}\ 15 = 20$

EXAMPLE | Use the given numbers to fill the blanks to make two *different* true statements.

Use: 4, 6, 10

$$\boxed{}-\boxed{}+\boxed{}=12$$

In the first two boxes, we look to subtract a smaller number from a larger one. This leaves three choices.

$6-4+10=12$ ✓
$10-6+4=8$ ✗
$10-4+6=12$ ✓

So, $\boxed{6}-\boxed{4}+\boxed{10}=12$

and $\boxed{10}-\boxed{4}+\boxed{6}=12$

PRACTICE | Use the given numbers to fill the blanks to make two *different* true statements.

44. **Use**: 17, 18, 19

$$\boxed{}-\boxed{}+\boxed{}=20 \quad\text{and}\quad \boxed{}-\boxed{}+\boxed{}=20$$

45. **Use**: 9, 19, 29

$$\boxed{}-\boxed{}-\boxed{}=1 \quad\text{and}\quad \boxed{}-\boxed{}-\boxed{}=1$$

46. **Use**: 11, 22, 44

$$\boxed{}-\boxed{}=\boxed{}+11 \quad\text{and}\quad \boxed{}-\boxed{}=\boxed{}+11$$

47. **Use**: 14, 16, 17

$$\boxed{}+15-\boxed{}=\boxed{} \quad\text{and}\quad \boxed{}+15-\boxed{}=\boxed{}$$

EXAMPLE | Evaluate $10-(6-4)$.

We first evaluate the part inside the parentheses, $6-4$. Then, we subtract the result from 10.

$$10-(6-4)$$
$$10-2$$
$$8$$

So, $10-(6-4)=\mathbf{8}$.

In an expression, anything inside parentheses gets evaluated first.

Within the parentheses, we add and subtract from left to right.

PRACTICE | Evaluate each expression below.

48. $(8-2)+6=$ _____

49. $8-(2+6)=$ _____

50. $8-2+6=$ _____

51. $9-(5-3)=$ _____

52. $(9-5)-3=$ _____

53. $9-5-3=$ _____

54. $(7+4)-1=$ _____

55. $7+(4-1)=$ _____

56. $7+4-1=$ _____

57. $13-12+(7-6)=$ _____

58. $13-(12+7-6)=$ _____

59. $28-(11-5)-2=$ _____

60. $28-(11-5-2)=$ _____

EXAMPLE | Place one pair of parentheses in the statement below to make it true.

$$12 - 6 - 3 + 2 = 11$$

Grouping the first two, three, or four numbers gives the same value as $12 - 6 - 3 + 2$:

$$(12 - 6) - 3 + 2 = 6 - 3 + 2 = 5$$
$$(12 - 6 - 3) + 2 = 3 + 2 = 5$$
$$(12 - 6 - 3 + 2) = 5$$

There are two other ways to use parentheses to group two numbers, and one other way to group three numbers:

$$12 - (6 - 3) + 2 = 12 - 3 + 2 = 11$$
$$12 - 6 - (3 + 2) = 12 - 6 - 5 = 1$$
$$12 - (6 - 3 + 2) = 12 - 5 = 7$$

From our choices, only $12 - \mathbf{(6 - 3)} + 2$ equals 11.

PRACTICE | Place one pair of parentheses in each statement below to make it true.

61. $15 - 6 + 4 = 5$

62. $20 - 4 - 3 = 19$

63. $25 - 3 + 2 + 5 = 25$

64. $70 - 50 - 30 - 10 = 0$

65. $45 - 18 - 17 + 9 = 35$

66. $10 - 2 + 3 - 1 = 4$

67. $100 - 45 + 25 + 15 = 45$

68. $17 + 6 - 5 + 4 = 14$

69. $7 + 6 - 5 + 4 - 3 = 1$

70. $15 - 4 + 6 - 3 + 2 = 10$

EXPRESSIONS
Parentheses

Expressions can have more than one pair of parentheses.

When there are parentheses inside parentheses, we evaluate the expression in the inner parentheses first.

EXAMPLE | Evaluate $20-(2+5)-(3+4)$.

$$20-(2+5)-(3+4)$$
$$= \quad 20-7-(3+4)$$
$$= \quad 20-7-7$$
$$= \quad 13-7$$
$$= \quad \mathbf{6}.$$

EXAMPLE | Evaluate $100-(50-(5+5))$.

$$100-(50-(5+5))$$
$$= \quad 100-(50-10)$$
$$= \quad 100-40$$
$$= \quad \mathbf{60}.$$

PRACTICE | Evaluate each expression below.

71. $10-(5-4)-(3+2) = $ _____

72. $1-(1-(1-1)) = $ _____

73. $30-3-(30-(10+3)) = $ _____

74. $24-(12-(3+3)-3) = $ _____

75. $39-(26-(7+9-5)) = $ _____

76. $4-(6-(8-(12-10))) = $ _____

PRACTICE | Circle the expression that can be used to answer each word problem below.

77. Alex has 12 red balloons and 7 green balloons. After 4 of the balloons pop, how many balloons does Alex have left?

$12-(7+4)$ $(12-7)-4$ $12-(7-4)$ $(12-7)+4$ $(12+7)-4$

78. Lizzie has 23 Beastbucks. Grogg had 38 Beastbucks, but spent 16 Beastbucks on a used hula hoop. After Grogg bought the hula hoop, how many Beastbucks do Lizzie and Grogg have together?

$23+(38-16)$ $38-(23+16)$ $38-(23-16)$ $23+38+16$ $(38-23)+16$

79. Ralph catches 25 centipillars. Cammie catches 15 centipillars, but 9 crawl away. How many more centipillars does Ralph have than Cammie?

$(25-9)-15$ $25-15-9$ $(25-15)-9$ $25-(15-9)$ $25+(15-9)$

80. Winnie uses a 64-ounce pitcher full of juice to fill four 12-ounce cups. How many ounces of juice are left in the pitcher?

$(64-4)-12$ $(64-12)+12+12+12$

$64-(12+12+12+12)$ $64-(4+12)$

PRACTICE | Answer each question below.

81. In the expression below, circle the + or − sign that should be replaced with the opposite sign to make the expression equal zero.

$$10 \; + \; 4 \; - \; 2 \; + \; 5 \; - \; 6 \; - \; 1$$

82. Circle the two expressions below that are equal.

6+12−1−5   12+1−5−6   12+5+1−6   5−1+6+12   12−6−5−1

83. What number is equal to nine minus the difference of four and two?

83. _____

84. ★ In a game of doubles beastball, Griff and Cliff played against Snark and Clark. Clark scored 11 points, Snark scored 19 points, Griff scored 15 points, and Cliff scored 21 points. Circle the expression below that shows how many points Griff and Cliff beat Snark and Clark by.

(15+21)−(19+11)   (21−15)+(19−11)

19−(15+(21−19))   21+11−(19−15)

PRACTICE | Answer each question below.

85. What is the **smallest** value you can get by placing one or
★ more pairs of parentheses in the expression below?

85. _____

$$16 - 8 - 4 + 2$$

86. What is the **largest** value you can get by placing one or
★ more pairs of parentheses in the expression below?

86. _____

$$16 - 8 - 4 + 2$$

87. Fill the blank below to make the statement true.
★

$$19 - (11 + (8 - \underline{\quad})) = 5$$

88. Fill each empty box below with +, −, or = to make a true statement.
★

$$4 \;\boxed{}\; 2 \;\boxed{}\; 1 \;\boxed{}\; 15 \;\boxed{}\; 3 \;\boxed{}\; 5$$

We can use symbols in expressions to stand for numbers.

EXAMPLE | If  = 4, what is ⬡−1+⬡?

To evaluate ⬡−1+⬡, we replace each symbol with the number it stands for. Then, we evaluate the expression.

This gives:

$$\begin{aligned}⬡-1+⬡ &= 4-1+4\\ &= 3+4\\ &= 7.\end{aligned}$$

PRACTICE | Evaluate each expression below.

89. Evaluate each expression below when ▲ = 3.

▲ + 2 = _____   ▲ − 1 = _____

4 + ▲ = _____   6 − ▲ = _____

90. Evaluate each expression below when ◆ = 9.

◆ + ◆ = _____   ◆ − 7 + ◆ = _____

20 − ◆ − 2 = _____   20 − (◆ − 5) − (◆ + 5) = _____

91. Evaluate each expression below when ■ = 99.

■ − 98 + ■ = _____   400 − ■ − ■ − ■ = _____

(■ + 2) − (■ − 1) = _____   ■ − 3 + ■ − 2 + ■ − 1 = _____

We can use more than one symbol in an expression.

EXAMPLE If ■ = 2 and ◆ = 5, what is ◆ + (◆ − ■)?

To evaluate ◆ + (◆ − ■), we replace each symbol with the number it stands for. Then, we evaluate the expression. This gives:

$$◆ + (◆ − ■) = 5 + (5 − 2)$$
$$= 5 + 3$$
$$= \textbf{8}.$$

PRACTICE | Evaluate each expression below when ☾ = 11 and ◯ = 6.

92. ☾ + 10 = _____

93. ◯ + ◯ + ◯ = _____

94. ☾ + ◯ = _____

95. ☾ − (◯ + 1) = _____

96. 18 − (◯ + ☾) = _____

97. ◯ + 3 − (☾ − 10) = _____

98. (☾ + ◯) − (☾ − ◯) = _____

99. ◯ + ☾ − (◯ + 5) = _____

EXAMPLE

Melinda buys 50−★ eggs at the market every Sunday.
The price in cents of one egg is ★.

How many eggs does she buy when eggs are 35 cents each?

How many eggs does she buy when eggs are 17 cents each?

When eggs cost 35 cents, ★ = 35.
So, 50−★ is 50−35 = 15, and Melinda buys **15** eggs.

When eggs cost 17 cents, ★ = 17.
So, 50−★ is 50−17 = 33, and Melinda buys **33** eggs.

PRACTICE | Solve each problem below.

100. Hope sells cups of lemonade for ✹−15 cents, where ✹ is the temperature in degrees. Find the price of a cup of lemonade for each temperature below.

45 degrees: _____ cents   68 degrees: _____ cents

85 degrees: _____ cents   99 degrees: _____ cents

101. Monsters with ✳ feet need to bring ✳+✳+✳+2 socks to ski camp. How many socks should each monster described below pack for ski camp?

4 feet: _____ socks   5 feet: _____ socks

10 feet: _____ socks   50 feet: _____ socks

PRACTICE | Solve each problem below.

102. Evaluate ▲−(10−◻) for each pair of values below.

▲ = 5, ◻ = 8: _____   ▲ = 25, ◻ = 3: _____

▲ = 2, ◻ = 9: _____   ▲ = 17, ◻ = 7: _____

103. Monsters in Mr. Baxter's class can earn "Baxter Bucks" for extra reading. Monsters who read ⬣ extra chapters in ⬥ weeks earn ⬣+⬣−⬥ Baxter Bucks. How many Baxter Bucks does each student below earn?

Trish reads 7 extra chapters in 2 weeks: _____ Baxter Bucks

Brian reads 3 extra chapters in 4 weeks: _____ Baxter Bucks

Dobbin reads 5 extra chapters in 10 weeks: _____ Baxter Bucks

104. In a game of Bobbleball, a team's score is ★+★+●−■, where ★ is the number of goals, ● is the number of saves, and ■ is the number of fouls. Find the team's score for each game below:

5 goals, 4 saves, 1 foul: _____ points

11 goals, 2 saves, 9 fouls: _____ points

0 goals, 6 saves, 4 fouls: _____ points

EXPRESSIONS

PRACTICE | Solve each problem below.

105. Robb makes double the number of donuts Globb makes. If Globb makes ⬤ donuts, which expression below describes the number of donuts **Robb** makes?

⬤   ⬤+2   ⬤+⬤+⬤   2   ⬤+⬤

106. Frida is 3 years older than Hana. If Hana is ⊢ years old, which expression below gives Frida's age in years?

⊢   ⊢−3   ⊢+3   ⊢+⊢+⊢   3−⊢

107. Captain Kraken has 100 coins. If he buys a helmet for ✸ coins, which expression below gives the number of coins he has left?

✸   100−✸   100+✸   ✸−(100−✸)   ✸−100

108. There are ★ players on the Speed Sledding team, ☾ of which are drivers. Which expression below gives the number of team members who are **not** drivers?

☾+★   ☾−★   ☾−(★−☾)   ★−☾   (★−☾)+☾

PRACTICE | Solve each problem below.

109. The number of girls in Ms. Franz's class is double the number of boys. If there are ● boys in the class, which expression below describes the ***total*** number of students in the class?

$$●+● \qquad ●-(●-●) \qquad ●+(●+●) \qquad ●+(●-●) \qquad ●-(●+●)$$

110. Tayvon is 5 inches taller than Stephanie. If Tayvon is T inches tall, which expression below gives Stephanie's height in inches?

$$T-5 \qquad 5-T \qquad T+5 \qquad T+(T-5) \qquad T-(T+5)$$

111. Every koalaphant has ▲ tusks. Which expression below gives the number of tusks in a group of 4 koalaphants?

$$▲-4 \qquad ▲+4 \qquad ▲+(▲-4) \qquad ▲+(▲+4) \qquad ▲+▲+▲+▲$$

112. ★ After Beastball practice, each team member does push-ups, sit-ups, and jumping jacks. Each team member must do a total of 50 exercises. Erp has already done S sit-ups and J jumping jacks. Which expression below gives the number of push-ups Erp must do?

$$50-S+J \qquad 50-(S-J) \qquad 50+(S+J) \qquad 50+(S-J) \qquad 50-(S+J)$$

Simplifying EXPRESSIONS

When we ***simplify*** an expression, we write it in a way that makes it easier to understand.

For example, we can simplify the expression 10+(999−998) to 10+1, which simplifies to 11.

We can also simplify many expressions that use symbols to stand for numbers.

EXAMPLE | Simplify ✦+532−532.

Adding 532 then taking away 532 is the same as doing nothing. So, if we start with ✦, add 532, then take away 532, we are left with ✦.

So, ✦+532−532 simplifies to ✦.

PRACTICE | Simplify each expression below.

113. 28−28 = ________

114. ♣−♣ = ________

115. 333−10+10 = ________

116. ■−45+45 = ________

117. 65−65+54 = ________

118. ♣−♣+11 = ________

119. 75−(57−57) = ________

120. ▲−(●−●) = ________

121. 19−(19−1) = ________

122. ★ ☾−(☾−3) = ________

EXAMPLE | Simplify ♣+1+2+3.

Adding 1, 2, and 3 is the same as adding 1+2+3 = 6. So, ♣+1+2+3 simplifies to ♣+6.

PRACTICE | Simplify each expression below.

123. ●+5+5+5+5

123. _____

124. ▲+(♣−♣)+(■−■)

124. _____

125. ◖−◖+◖+7

125. _____

126. 18+☾+32

126. _____

127. 20−⭘−10
★

127. _____

128. (▼+5)−(▼+2)
★

128. _____

129. ◨−(◨−◆)
★
★

129. _____

Simplifying an expression can make evaluating a lot easier.

PRACTICE | Solve each problem below.

130. Evaluate $\blacktriangledown - 19 + 19 - 19$ for each value of $\blacktriangledown$ below.

$\blacktriangledown = 20:$ _____

$\blacktriangledown = 25:$ _____

$\blacktriangledown = 49:$ _____

$\blacktriangledown = 87:$ _____

131. Evaluate $100 - \bigstar - \bigstar + \bigstar$ for each value of $\bigstar$ below.

$\bigstar = 15:$ _____

$\bigstar = 5:$ _____

$\bigstar = 33:$ _____

$\bigstar = 49:$ _____

132. Evaluate $\maltese + (\bigcirc - \bigcirc) + (\bigcirc - \bigcirc) + \bigcirc$ for each value of $\maltese$ and $\bigcirc$ below.

$\maltese = 6, \bigcirc = 8:$ _____

$\maltese = 20, \bigcirc = 10:$ _____

$\maltese = 33, \bigcirc = 27:$ _____

$\maltese = 111, \bigcirc = 86:$ _____

133. Evaluate $15 + (\blacksquare - \leftmoon) - (\blacksquare - \leftmoon) + \blacksquare$ for each value of $\blacksquare$ and $\leftmoon$ below.
$\bigstar$

$\blacksquare = 61, \leftmoon = 48:$ _____

$\blacksquare = 143, \leftmoon = 74:$ _____

$\blacksquare = 212, \leftmoon = 188:$ _____

$\blacksquare = 800, \leftmoon = 651:$ _____

Equations show that two expressions are equal.

In an **equation**, two expressions are separated by an equals sign (=).

The expression on the left side of the equals sign is equal to the expression on the right side.

Below are some examples of equations.

$$5+6=11$$
$$20+20+20=30+30$$
$$135-99=135-100+1$$
$$59+59+59=60+60+60-3$$

left side ⟷ right side

PRACTICE | Fill each ◯ with +, −, or = to make an equation.
For example, 5 ⊜ 7 ⊖ 2.

134. 1 ⊕ 2 ⊜ 3

135. 1 ⊕ 3 ⊜ 5 ⊖ 1

136. 4 ⊕ 5 ⊕ 6 ⊜ 15

137. 1 ⊜ 3 ⊕ 2 ⊖ 4

138. 20 ⊜ 40 ⊖ 15 ⊖ 5

139. 25 ⊖ 24 ⊕ 26 ⊜ 27

140. 16 ⊜ 18 ⊖ 14 ⊕ 12

141. 3 ⊕ 1 ⊜ 8 ⊖ 1 ⊖ 3

In a **Mismo** puzzle, the goal is to fill each empty circle with +, −, or = to create equations that are read from top-to-bottom and from left-to-right.

EXAMPLE | Complete the Mismo puzzle on the right.

$$9 \quad 5$$
$$7 \bigcirc 4 \bigcirc 3$$
$$8 \quad 7$$
$$6 \bigcirc 10 \bigcirc 4$$
$$1 \quad 2$$

There are two ways to fill the top two circles to make an equation in the top row: 7 (=) 4 (+) 3 or 7 (−) 4 (=) 3.

If we fill the top-right circle with +, then we cannot create an equation reading down from five: 5 (+) 7 (◯) 2.

So, we use 7 (−) 4 (=) 3 and complete the puzzle as shown.

$$9 \quad 5$$
$$7 (-) 4 (=) 3$$
$$8 \quad 7$$
$$6 (=) 10 (-) 4$$
$$1 \quad 2$$

PRACTICE | Complete each Mismo puzzle below.

Find more Mismo puzzles at BeastAcademy.com.

142.

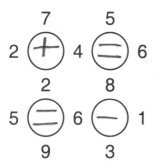

$$7 \quad 5$$
$$2 (+) 4 (=) 6$$
$$2 \quad 8$$
$$5 (=) 6 (-) 1$$
$$9 \quad 3$$

143.

$$6 \quad 13$$
$$9 (-) 15 (-) 6$$
$$13 \quad 4$$
$$15 (-) 7 (=) 8$$
$$7 \quad 9$$

PRACTICE | Complete each Mismo puzzle below.

144.

```
       4        3
   8 ( = ) 7 ( + ) 1
      10        9
   5 ( - ) 3 ( = ) 2
       6       12
```

145.

```
       2        6
   2 ( + ) 8 ( = ) 10
       6        4
  12 ( = ) 10 ( + ) 2
       8        2
```

146.

```
      15       16
   6 ( = ) 5 ( = ) 1
       9        8
  11 ( = ) 3 ( + ) 8
       6        8
```

147.

```
       9        9
   9 ( - ) 1 ( = ) 8
       4        3
   9 ( = ) 2 ( + ) 7
       5        6
```

148.

```
       3        3        3
   1 ( + ) 1 ( + ) 1 ( = ) 3
       3        4        5
   6 ( + ) 1 ( = ) 9 ( - ) 2
       3        8        1
   7 ( = ) 9 ( - ) 1 ( - ) 1
       9        1        1
```

149.

```
      21       21       21
   3 ( + ) 12 ( - ) 5 (   ) 4
       5        7        6
   6 ( + ) 4 ( = ) 15 ( - ) 5
       7        7        7
   7 (   ) 6 ( + ) 5 (   ) 18
       9        7        8
```

PRACTICE | Complete each Mismo puzzle below.

150.

```
      1       5       11
 20 ( )  1 ( ) 22 ( )  1
      1      23       2
 12 ( ) 17 ( )  7 ( )  2
      3      22       9
  8 ( )  4 ( )  5 ( )  9
      1       4       18
```

151.

```
      2       1       4
  4 ( )  2 ( )  1 ( )  3
      1       2       2
  1 ( )  2 ( )  4 ( )  1
      3       4       1
  2 ( )  3 ( )  2 ( )  1
      4       1       3
```

152.

```
      6       3       9
  5 ( )  4 ( )  2 ( )  3
      1       6       8
  7 ( )  8 ( )  6 ( )  5
      2       5       7
  2 ( )  3 ( )  9 ( )  4
      3       4       6
```

153.

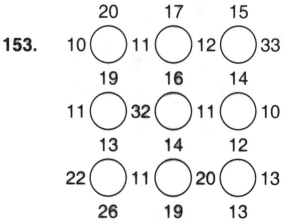

```
     20      17      15
 10 ( ) 11 ( ) 12 ( ) 33
     19      16      14
 11 ( ) 32 ( ) 11 ( ) 10
     13      14      12
 22 ( ) 11 ( ) 20 ( ) 13
     26      19      13
```

154. ★

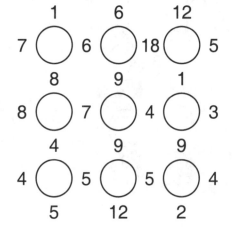

```
      1       6       12
  7 ( )  6 ( ) 18 ( )  5
      8       9       1
  8 ( )  7 ( )  4 ( )  3
      4       9       9
  4 ( )  5 ( )  5 ( )  4
      5      12       2
```

155. ★ ★

```
     15      15       5
 15 ( ) 15 ( )  5 ( )  5
     25      15       5
 15 ( ) 15 ( )  5 ( )  5
     15       5      25
 15 ( )  5 ( ) 25 ( )  5
      5       5       15
```

Finding the value of a symbol in an equation is called **solving** the equation.

EXAMPLE | If ⬡+⬡+7 = 19, what is ⬡?

Since <u>12</u>+7 = 19, we know that ⬡+⬡ must equal 12.

$$⬡+⬡+7 = 19$$
$$12 \ \ +7 = 19$$

Since 6+6 = 12, we know that ⬡ = **6**.

We replace ⬡ with 6 to check our answer:

$$6+6+7 = 19. \ ✓$$

PRACTICE | Find the value of the symbol in each equation below.

156. ⬖+5 = 11   ⬖ = _____

157. 12+✶ = 20   ✶ = _____

158. ★−40 = 10   ★ = _____

159. 16 = ◯−100   ◯ = _____

160. 22 = ●+●   ● = _____

161. ✳+✳+1 = 51   ✳ = _____

162. ★ 24−◆ = ◆   ◆ = _____

163. ★ 5+▲ = ▲+▲   ▲ = _____

164. ★ 13−(☾+3) = 5   ☾ = _____

165. ★ 24−♣ = ♣+♣   ♣ = _____

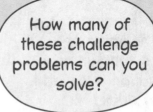

How many of these challenge problems can you solve?

PRACTICE | Answer each question below.

166. Fill the blanks with the six expressions below to make three equations.

$$5+5+5 \qquad 5+4-3 \qquad 3+4-5 \qquad 2+2+2 \qquad 3-2+1 \qquad 4+5+6$$

_____ = _____

_____ = _____

_____ = _____

167. Fill each ◯ with +, −, or = to create three *different* equations.
★

101 ◯ 1 ◯ 99 ◯ 1

101 ◯ 1 ◯ 99 ◯ 1

101 ◯ 1 ◯ 99 ◯ 1

PRACTICE | Answer each question below.

168. Place *one pair* of parentheses in the statement below to make it true.
★

$$32 - 16 - 8 - 4 - 2 - 1 = 13$$

169. Find two different ways to place *two pairs* of parentheses in the statement
★ below to make it true.

$$5 - 4 - 3 - 2 - 1 = 3$$

$$5 - 4 - 3 - 2 - 1 = 3$$

170. Simplify the expression $(☾+10)-(☾-10)$.
★

170. _____

171. If $▲+★=▲+17$ and $▲+▲+★=▲+39$,
★ what is $▲$?
★

171. $▲=$ _____

CHAPTER 6
Problem Solving

This is a very challenging chapter that will help you learn some great ways to solve tough problems.

Take your time, and don't worry if you can't solve many of the problems on your first try!

Use this Practice book with Guide 2B from BeastAcademy.com.

Recommended Sequence:

Book	Pages:
Guide:	76-79
Practice:	73-81
Guide:	80-84
Practice:	82-89
Guide:	85-101
Practice:	90-107

You may also read the entire chapter in the Guide before beginning the Practice chapter.

When you don't know how to solve a problem, sometimes you can learn a lot by **guessing**, then checking your answer.

EXAMPLE | Trent adds three consecutive numbers and gets 51. What is the smallest of these three numbers?

Consecutive numbers come one after the other. We try guessing.

Guess: The smallest number is 10. Then the numbers are 10, 11, and 12, and their sum is $10+11+12=33$. Too small.

Guess: The smallest number is 20. Then the numbers are 20, 21, and 22, and their sum is $20+21+22=63$. Too big.

Guess: The smallest number is 15. Then the numbers are 15, 16, and 17, and their sum is $15+16+17=48$. Almost! A little too small.

Guess: The smallest number is 16. Then the numbers are 16, 17, and 18, and their sum is $16+17+18=51$. Got it!

So, the smallest number is **16**.

PRACTICE | Fill in the blanks to solve the problem in bold below.

Tram is 5 years older than Sam. The sum of their ages is 31. How many years old is Sam?

1. We guess that Sam is 10. Since Tram is 5 years older,

    Tram's age would be _____. The sum of their ages would be _____.

2. We guess that Sam is 15. Since Tram is 5 years older,

    Tram's age would be _____. The sum of their ages would be _____.

3. Is Sam older or younger than 15?

    3. _____

4. How many years old is Sam?

    4. _____

PRACTICE | Fill the blanks to solve each problem given in bold below.

Ming has twice as many toes as Ned. Together, they have 54 toes. How many toes does Ned have?

5.   We guess that Ned has 15 toes. Since Ming has twice as many, she would

have _____ toes. Together they would have a total of _____ toes.

6.   We guess that Ned has 20 toes. Since Ming has twice as many, she would

have _____ toes. Together they would have a total of _____ toes.

7.   How many toes does Ned have?   7. _____

Mike has $5 bills and $10 bills worth a total of $55. Mike has a total of 9 bills. How many $5 bills does Mike have?

8.   We guess that Mike has four $5 bills. Then, he would need _____ $10 bills to

have a total of 9 bills. The total value of Mike's bills would be _____ dollars.

9.   We guess that Mike has six $5 bills. Then, he would need _____ $10 bills to

have a total of 9 bills. The total value of Mike's bills would be _____ dollars.

10.   How many $5 bills does Mike have?   10. _____

PRACTICE | In each equation below, fill every blank with the **same number**.

11. ____ + ____ + ____ = 30

12. ____ + ____ + ____ + ____ = 52

13. 78 − ____ = ____

14. ★ ___ + 16 = 54 − ____

15. ★ ____ + ____ = 27 − ____

16. ★ 30 + ____ = 50 − (4 + ____)

PRACTICE | Solve each problem below.

17. The difference between two numbers is 5.
The sum of the two numbers is 13.
What is the larger of the two numbers?

17. _____

18. Brenda has 6 coins, all nickels and dimes.
Nickels are 5 cents and dimes are 10 cents.
Brenda's 6 coins are worth 40 cents all together.
How many nickels does Brenda have?

18. _____

19. ★ Chickens have 2 legs and cows have 4 legs. Mr. Hayfield
keeps only chickens and cows on his farm. If his 10
animals have 26 legs, how many cows does he have?

19. _____

PRACTICE | Solve each problem below.

20. ★ Grogg picks a two-digit number. Winnie's number is the sum of the two digits of Grogg's number. Grogg's number is double Winnie's number. What is Grogg's number?

20. _____

21. ★ If △ + △ + ▢ = 13 and △ + ▢ + ▢ = 17, then what is the value of △ + ▢?

21. △ + ▢ = _____

22. ★ A small bag and a big bag have 42 apples all together. The big bag has twice as many apples as the small bag. How many apples must be moved from the big bag to the small bag so that both bags have the same number of apples?

22. _____

In a **Sym-Sum** puzzle, each symbol stands for a whole number. The numbers below and to the right of the grid give the sum of the symbols in each column and row. The goal is to find the value of each symbol.

EXAMPLE | Find the value of △, ○, and □ in the Sym-Sum puzzle to the right.

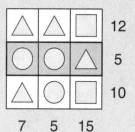

In the middle row, we see that ○+○+△ = 5.
Since 3+3 = 6, we know that ○ must be less than 3.
We try to guess the value of ○.

Guess: ○ = 2.
Since the sum of the middle row is 5, we have △ = 1.
If ○ = 2 and △ = 1, then the left column would have a sum of 1+2+1 = 4. But, the left column has a sum of 7, not 4. So, ○ cannot be 2.

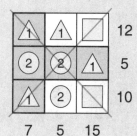

Guess: ○ = 1.
Since the sum of the middle row is 5, we have △ = 3.
If ○ = 1 and △ = 3, then the left column would have a sum of 3+1+3 = 7. This works!

So, we use ○ = 1 and △ = 3 to find the value of □.

In the bottom row, we have △+○+□ = 10.
Replacing △ with 3 and ○ with 1, we have
3+1+□ = 10.

Since 3+1+6 = 10, we have □ = 6.

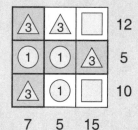

We check the other rows and columns to make sure each sum is correct. So, △ = **3**, ○ = **1**, and □ = **6**.

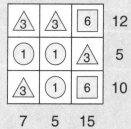

PRACTICE | Find the value of △, ○, and □ in each Sym-Sum puzzle below.

23.

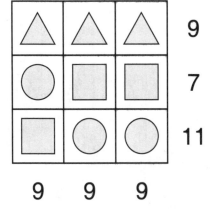

△ = ____, ○ = ____, □ = ____

24.

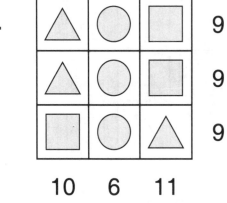

△ = ____, ○ = ____, □ = ____

25.

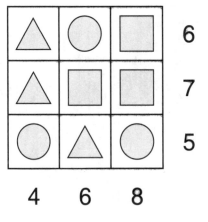

△ = ____, ○ = ____, □ = ____

26.

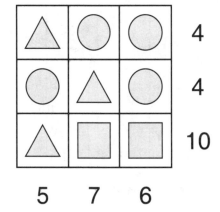

△ = ____, ○ = ____, □ = ____

PRACTICE | Find the value of △, ◯, and ▢ in each Sym-Sum puzzle below.

27.

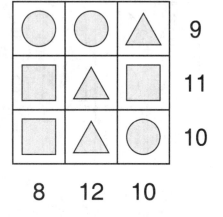

△ = _____, ◯ = _____, ▢ = _____

28.

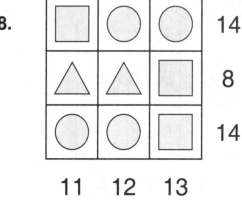

△ = _____, ◯ = _____, ▢ = _____

29.

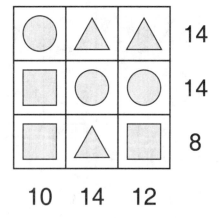

△ = _____, ◯ = _____, ▢ = _____

30.

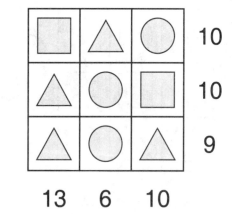

△ = _____, ◯ = _____, ▢ = _____

PRACTICE | Find the value of △, ◯, and ▢ in each Sym-Sum puzzle below.

31.

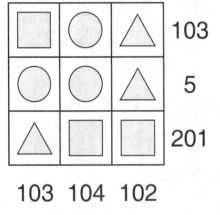

△ = ____, ◯ = ____, ▢ = ____

32.

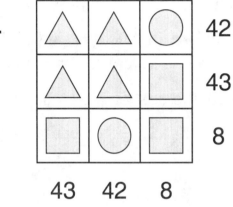

△ = ____, ◯ = ____, ▢ = ____

33.
★

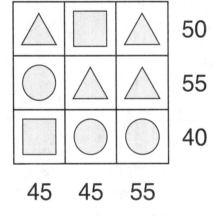

△ = ____, ◯ = ____, ▢ = ____

34.
★

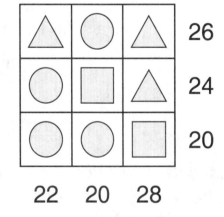

△ = ____, ◯ = ____, ▢ = ____

EXAMPLE

A bus leaves the station and drops off 25 passengers at its first stop. It picks up 8 passengers at its second stop. All 39 passengers get off at the third stop. How many passengers were on the bus when it left the station?

We work backwards. Since all 39 passengers got off at the third stop, *before* the third stop, there were 39 passengers on the bus.

Before the bus picked up 8 passengers at the second stop, there were $39-8=31$ passengers on the bus.

Before the bus dropped off 25 passengers at the first stop, there were $31+25=56$ passengers on the bus.

So, the bus left the station with **56** passengers. We check our work.

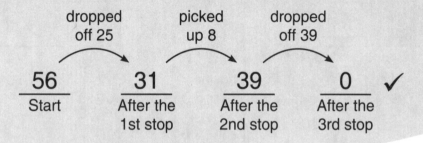

dropped off 25		picked up 8		dropped off 39	

$\underline{\quad 56 \quad}$ $\underline{\quad 31 \quad}$ $\underline{\quad 39 \quad}$ $\underline{\quad 0 \quad}$ ✓
Start   After the 1st stop   After the 2nd stop   After the 3rd stop

PRACTICE | Answer each question below.

35. Ted earned 12 dollars doing chores Friday afternoon, then spent 20 dollars on a video game. Ted has 45 dollars left. How many dollars did Ted have before doing chores?

35. _____

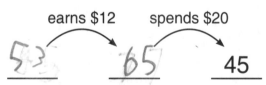

earns $12   spends $20

$\underline{\quad 53 \quad}$ $\underline{\quad 65 \quad}$ $\underline{\quad 45 \quad}$

36. Mira bakes a batch of cookies to sell at a school fundraiser. She sells 25 of her cookies, buys 12 cookies from another parent, eats 3 of them, and brings a total of 18 cookies home. How many cookies did Mira bake?

36. _____

sells 25   buys 12   eats 3

$\underline{\quad 34 \quad}$ $\underline{\quad 9 \quad}$ $\underline{\quad 21 \quad}$ $\underline{\quad 18 \quad}$

PRACTICE | Answer each question below.

37. Hildy is 5 inches taller than Marge.
Marge is 8 inches shorter than Sheila.
Sheila is 2 inches taller than Betsy.
Betsy is 45 inches tall. How many inches tall is Hildy?

37. $\underline{44}$

38. Hector leaves his hotel room and gets on the elevator. The elevator goes up 6 floors, down 18 floors, up 9 floors, then up 15 more floors, where Hector steps off onto the 36th floor to visit the hotel spa. What floor is Hector's room on?

38. $\underline{24}$

39. Completing each level of Beast Blasters is worth twice as many points as completing the previous level. Completing level 4 is worth 600 points. How many points do players earn for completing level 1?

39. $\underline{\hspace{2cm}}$

40. ★ Skippy Squirrel collects his acorns in a hole. Every day, he collects 12 acorns, eats 5, and gives one to his friend, Chippy Chipmunk. At the end of the day on Friday, Skippy has 100 acorns. How many acorns did Skippy have at the end of the day on Monday, four days earlier?

40. $\underline{\hspace{2cm}}$

A **modibot** can increase or decrease a number you put into it. Below are some examples of modibots.

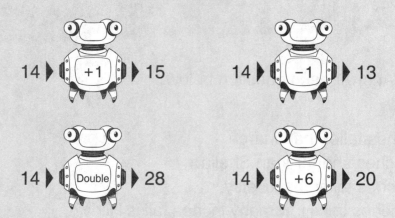

Connecting two or more modibots creates a **combobot**. For example, the combobot below doubles a number, then subtracts 1. So, if you give it 14, the first bot doubles this to get $14+14=28$, and the second bot subtracts 1, leaving $28-1=27$.

PRACTICE | Fill the blank to solve each problem below.

41.  22

42.  16

43. 29 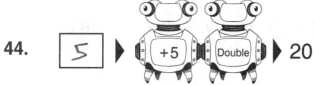 35

44. 5 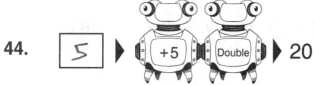 20

PRACTICE | Fill the blank to solve each problem below.

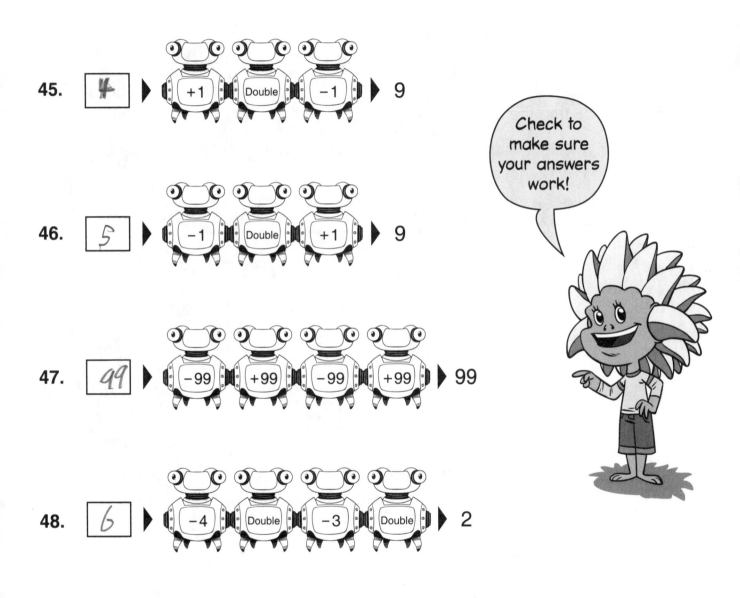

45. **4** ▶ +1 → Double → −1 ▶ 9

Check to make sure your answers work!

46. **5** ▶ −1 → Double → +1 ▶ 9

47. **99** ▶ −99 → +99 → −99 → +99 ▶ 99

48. **6** ▶ −4 → Double → −3 → Double ▶ 2

49. **13** ▶ Double → −9 → Double → −9 → Double ▶ 50

In the game of chess, the **knight** moves in an L shape: 2 squares in one direction, then 1 square to either side (or 1 square in one direction, then 2 squares to either side). For example, the knight on the chess board below could move to any one of the 8 squares shown.

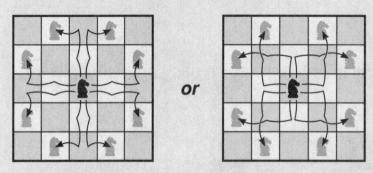

or

In a **Knight Path** puzzle, the goal is to number some spaces on a chess board, in order, so that the knight lands on each numbered space exactly once.

Below is a Knight Path puzzle and its solution. We fill the first circle the knight moves to with 1, the second circle it moves to with 2, and so on until all seven circles have been filled.

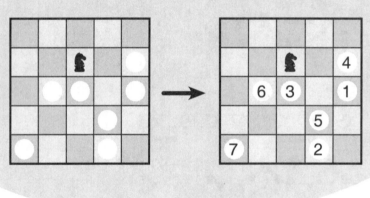

How can working backwards make these easier?

PRACTICE | Label the empty circles below from 1 to 4 so that the knight lands on each space exactly once before landing on the 5.

50.

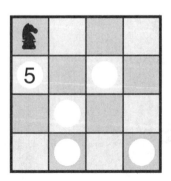

51.

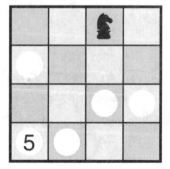

PRACTICE | Label the empty circles below in order so that the knight lands on each space exactly once.

52.

53.

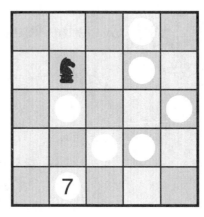

54.
★

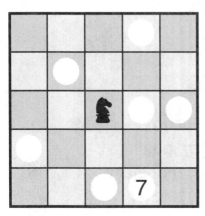

55.
★

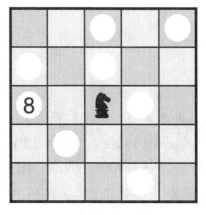

56.
★
★

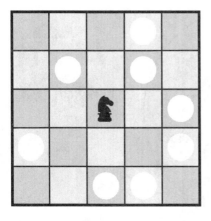

57.
★
★

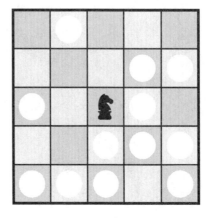

PRACTICE | Answer each question below.

58. Circle the letter whose ● is connected by a string to the Finish ●.

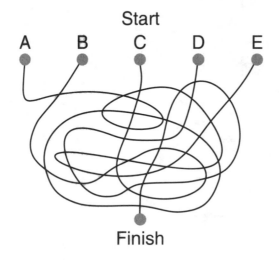

59. ★ Art, Burt, Curt, Dart, Ert, Fort, and Gert are all sitting in a circle. Zog walks around the circle as shown below, putting a sticker on every **third** monster. Curt was the last monster to get a sticker. Who was the first monster to get a sticker?

59. _____

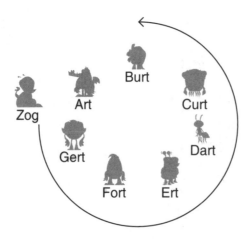

PRACTICE | Answer each question below.

60. Lizzie is practicing addition with a game. She starts by writing three numbers
★
★ on a piece of paper. Each turn, she crosses out one number and adds the
other two, writing the sum somewhere on her paper. After four turns, Lizzie's
paper looks like this:

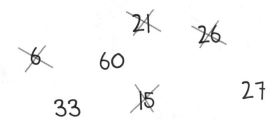

What three numbers did Lizzie start with? ____, ____, and ____

61. In a DigiSum list, we find each number using the two numbers that come
before it. We add the last number plus the *digits* of the number before it to get
the next number, as shown below.

$$24, \ 31, \ \underline{37}$$

Fill in the missing numbers in each DigiSum list below.

a. 15, 18, 24, 33, ____, ____, ____, ____

b. 12, ____, ____, ____, ____, 29, 38, 49
★
★

EXAMPLE | Amy and Mel are in line. Amy is 17th in line and Mel is 27th. How many monsters are between them?

We might think that since 27 − 17 = 10, there are 10 monsters between Amy and Mel. To check, we draw a diagram, as shown below.

Amy
17 (18) (19) (20) (21) (22) (23) (24) (25) (26) Mel
                                                    27

From our picture, we count that there are actually **9** monsters between Amy and Mel.

> Simple drawings are a great way to organize information.

PRACTICE | Use the picture to help you answer each question below.

62. There are eight doors along a wall, with one light between each pair of doors. How many lights are there?

62. ___7 lights___

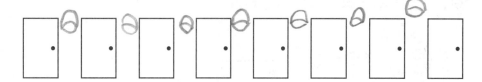

63. A square field is surrounded by a fence. The field has five fence posts on each side, counting the posts at the corners. How many fence posts are around the field?

63. ___16 posts___

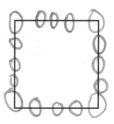

PRACTICE | Draw a picture to help you answer each question below.

64. Monsters are seated in the front row of the Beast Academy auditorium so that every boy is seated between two girls. There are 7 boys in the front row. What is the smallest number of girls that could be in the front row?

64.  8 girls

GBGBGBGBGBGBGBG

65. Bronck is jogging around a square block. Beginning at one corner, he jogs two full laps around the block, making only right turns. He finishes at the corner where he started. How many right turns did Bronck make?

65. 7 turns

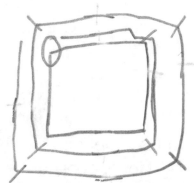

66. Alex makes a circle of coins using 3 nickels, 3 dimes, and some pennies. None of the coins that are next to each other are the same. What is the largest number of pennies Alex could have placed in the circle?

66.  6 pennies

Draw a picture to solve these problems!

PRACTICE | Answer each question below.

67. A group of kids line up in a row and hold hands. If there are 12 hand-holds, how many kids are there?

67. _____

68. ★ Tarrin the termite lives in a tree. Lenny the locust lives in the same tree, 24 feet below Tarrin. Lenny and Tarrin meet on a branch 100 feet above the ground. The branch is 56 feet above Lenny's home. How many feet above the ground is Tarrin's home?

68. _____

69. ★ Ben is standing 5 feet from Amy, who is 9 feet from Kenneth, who is 6 feet from Deepa. All four people are standing in a line. What is the smallest number of feet that Ben could be from Deepa?

69. _____

PRACTICE | Answer each question below.

70. Kiana lays out a row of colored blocks in a straight line. She starts with a row of 5 red blocks. Between every pair of red blocks, she places a blue block. Then, she puts a green block between every neighboring red and blue block. How many green blocks are in Kiana's row?

70. _____

71. ★ Grammy Rosie baked 10 cookies. She put chocolate chips in 6 of them and pecans in 7 of them. What is the smallest number of cookies that could have both chocolate chips and pecans?

71. _____

72. ★ The town of Bantam has just four straight roads. Wherever two roads cross, there is one stoplight. What is the largest number of stoplights that can be in Bantam?

72. _____

EXAMPLE | Four people stand in line. Kat is third in line. Larry is behind Mark but in front of Norm. Who is last in line?

We draw blanks for the 1st, 2nd, 3rd, and 4th spots in line.
Kat is 3rd in line, so we draw a K for Kat in the 3rd spot.

$$\frac{\quad}{1^{st}} \quad \frac{\quad}{2^{nd}} \quad \frac{K}{3^{rd}} \quad \frac{\quad}{4^{th}}$$

Then, there is only one place Larry (L) can stand so that he is behind Mark (M) and in front of Norm (N), as shown below.

$$\frac{M}{1^{st}} \quad \frac{L}{2^{nd}} \quad \frac{K}{3^{rd}} \quad \frac{N}{4^{th}}$$

So, **Norm** is last in line.

PRACTICE | Answer each question below.

73. Draw a ■, a ◆, a ●, and a ▲ in the blanks below using the following rules:
    • The ■ is to the right of the ◆ but to the left of the ●.
    • The ▲ is between the ■ and the ●.

    _____   _____   _____   _____

74. Kim, Tim, Jim, and Finn are comparing heights. Kim is taller than Tim, but shorter than Jim. Finn is taller than Tim, but shorter than Kim. Write the names of the four friends in order below from **shortest** to **tallest**.

    _____   _____   _____   _____

PRACTICE | Answer each question below.

75. Beast Theaters offers four sizes of popcorn tubs.
The red tub is smaller than the blue tub.
The yellow tub is bigger than the green tub.
The green tub is bigger than the blue tub.
What color tub is the smallest?

75. _____

76. Tim writes a word using the letters O, P, S, and T once each.
The letter T comes after P and before S. The letter O is not
the first or the last letter. What word did Tim write?

76. _____

77. ★ Flynn finished 3rd in a race. He beat twice as many
monsters as Wynn, who came in 7th. How many monsters
were in the race?

77. _____

In a **SumBox** puzzle, we arrange boxes so that the numbers that are next to each other always have a sum that ends in one of the target digits.

EXAMPLE | Solve the SumBox puzzle on the right.

Target digits: 4, 7

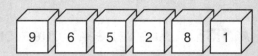

For each number, we find the numbers it can be next to. For example, since $9+5=1\underline{4}$ and $9+8=1\underline{7}$, the 9 can be next to 5 or 8. We list the possibilities for all six numbers below.

The 9 can be next to 5 or 8.

The 6 can be next to 1 or 8.

The 5 can be next to 9 or 2.

The 2 can only be next to 5.

The 8 can be next to 6 or 9.

The 1 can only be next to 6.

This information can be shown in a drawing like the one on the right. We connect numbers that can be next to each other.

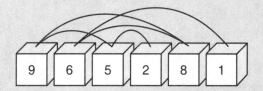

In our drawing, we see that 1 can only be next to 6, which must be next to 8, then 9, then 5, and finally 2. So, these boxes could be arranged in either order below.

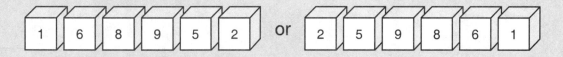

PRACTICE | Solve each SumBox puzzle below.

78. **Target digits**: 2, 5

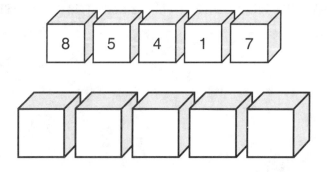

PRACTICE | Solve each SumBox puzzle below.

79. **Target digits**: 2, 9

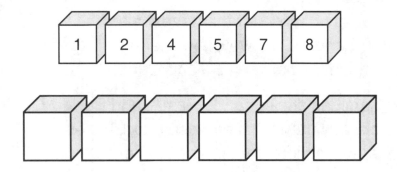

80. **Target digits**: 1, 8

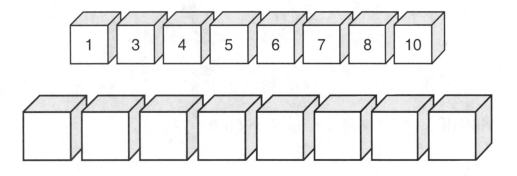

81. **Target digits**: 3, 6
★

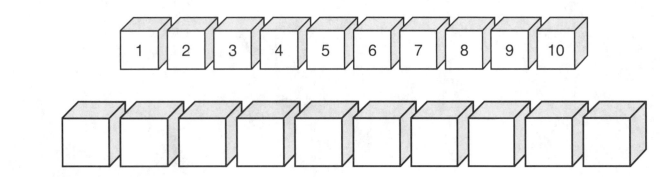

EXAMPLE | Al, Bo, and Cam meet, and everyone shakes hands. How many different handshakes are possible among all three monsters?

Drawing can help us organize the handshakes. We draw a dot for Al, Bo, and Cam. Then, we connect two dots to stand for each handshake.

Al shakes hands with Bo and Cam.
So, we connect Al to Bo and to Cam.

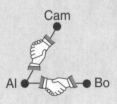

Bo can shake hands with Al and with Cam. We already drew a line for the handshake between Bo and Al. So, we draw one more line for the handshake between Bo and Cam.

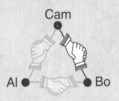

Finally, Cam can shake hands with Al and Bo. But, we already drew both of these handshakes. So, there are no more handshakes to consider.

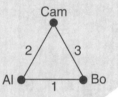

All together, **3** different handshakes are possible among Al, Bo, and Cam.

PRACTICE | Answer each question below.

82. In the example above, we saw that 3 different handshakes are possible in a group of 3 monsters. How many different handshakes are possible in a group of 4 monsters?

82. _____

PRACTICE | Answer each question below.

83. How many different handshakes are possible in a group of 5 monsters?

83. _____

84. The Hatfield triplets and the McCoy twins meet at the park. All five monsters shake hands with each other, except for Matty McCoy, who refuses to shake hands with any of the Hatfield triplets. How many handshakes are there?

84. _____

85. ★ A group of 4 monsters meet. Fred shakes 3 monsters' hands, Gary shakes 2 monsters' hands, and Holden shakes 1 monster's hand. How many monsters did Iggy shake hands with?

85. _____

The game of **Match** is played with a special set of 27 cards.

Each card has three *features*: a shape (○, △, or □), a size (small, medium, or large), and a color (white, gray, or black). Below are some sample cards.

This card is a medium white circle.

Players are dealt 5 cards. The rest are placed face-down in a **draw pile**. The top card from the draw pile is turned face-up to begin a new **stack**. Players then take turns placing one card on the stack. You can only place a card on the stack if it matches *two* features of the card on the top of the stack. For example, six different cards can be played on a small black triangle:

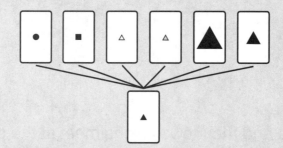

If you cannot play a card on your turn, draw one card from the draw pile and end your turn.

The first person to play all of the cards in their hand is the winner.
If the draw pile runs out, the player with the fewest cards in their hand wins.

For rule clarifications, variations, and card cut-outs, visit BeastAcademy.com

PRACTICE | For each problem below, circle *all* of the cards on the right that can be played on the given card on the left.

86. |

87. |

PRACTICE | Solve each Match problem below.

88. The stack in a game of Match contains the five cards below.
The **large black square** is on top of the stack. Circle the card on the bottom.

89. The stack in a game of Match contains the five cards below.
The **large white circle** is on top of the stack. Circle the card on the bottom.

90. The stack in a game of Match contains the five cards below.
The **medium gray square** is on top of the stack. Circle the card on the bottom.

91. The stack in a game of Match contains the six cards below.
★ The **small black square** is on top of the stack. Circle the card on the bottom.

EXAMPLE

There are 8 red balls, 7 green balls, and 14 yellow balls in a bag. Anna removes all but 2 of the green balls. How many green balls did Anna remove?

Some problems give more information than you need.

There are 7 green balls, and Anna removes all but 2 of them.

So, Anna removed 7−2 = **5** green balls.

The number of red and yellow balls did not matter.

You will need to find the useful information in these problems.

PRACTICE | Answer each question below.

92. At 12:15 PM, Mike took 7 cookies from the cookie jar. He ate 3 of them and put the rest in his pocket for later. The cookie jar now has 20 cookies in it. How many cookies were in the jar before Mike took some out?

92. _27_

$20 + 7$

93. Three years ago, on Jane's 6th birthday, she was 38 inches tall and weighed 39 pounds. Since then, she has grown 20 inches and gained 60 pounds. How many more years will it be before Jane turns 16?

93. _7 years_

PRACTICE | Answer each question below.

94. A box holds 15 pairs of shoes: 4 boots, 8 sandals, 10 sneakers, 2 loafers, and the rest are high heels. How many shoes are in the box?

94. _____

95. A class of 33 students is seated in three rows. The rows have 7, 11, and 15 students. Each row has one more boy than it has girls. How many more boys are there than girls in the class?

95. _____

96. ★ Grogg flips 18 coins: 5 dimes, 6 nickels, and 7 pennies. The coins that land heads are worth a total of 33 cents. How many pennies landed heads?

96. _____

97. ★ Four students each answer 20 questions. The four students get 14, 15, 16, and 17 answers correct. Jeremy gets twice as many answers wrong as Melissa. Melvin gets one more answer correct than Yasmin. How many questions did Melissa answer correctly?

97. _____

Try out some of the strategies you've learned!

PRACTICE | Answer each question below.

98. Draw three straight cuts on the pizza to the right that split it into 7 different pieces.

99. Two consecutive numbers have a sum of 387. What is the difference between the two numbers?

99. _____

100. Al, Lynn, Jon, Nick, Nora, and Lionel stand in line so that the last letter in each person's name is the first letter of the person standing behind them. If Nick is at the back of the line, who is in the front of the line?

100. _____

PRACTICE | Answer each question below.

101. The sum of three different whole numbers is 37. The smallest number is 11. What is the largest number?

101. _____

102. ★ Shelly Snail is climbing a slippery wall. She can glide 3 feet up the wall, but always slips down 2 feet while she rests before her next glide. How many glides does it take for Shelly to reach the top of the 6-foot wall?

102. _____

103. ★ Alex writes the expression shown below.

$$(4+6)+(7+8)$$

Winnie writes the same expression as Alex, but switches the 4 and the 6. Grogg writes the same expression as Winnie, but switches the 7 and the 8. Lizzie writes the same expression as Grogg, but switches the 6 and the 7. What is the value of Lizzie's expression?

103. _____

PRACTICE | Answer each question below.

104. The three Pajama islands are connected by bridges.　　　**104.** _____
★　There are 5 bridges from Llama Island to the other two islands.
There are 4 bridges from Mama Island to the other two islands.
There are 3 bridges from Nama Island to the other two islands.
How many bridges connect Mama Island and Nama Island?

105. Jorm has fewer than 10 PokeBeast cards. He goes to the　　**105.** _____
★　store and buys cards in packs of 8. Now, Jorm has 61 cards.
How many cards did Jorm have before going to the store?

106. Alicia, Betsy, Carla, and Denise are all different ages.　　**106.** _____
★　Carla and Betsy are 1 year apart.
Alicia and Denise are 2 years apart.
Alicia and Carla are 3 years apart.
Alicia and Betsy are 4 years apart.
Carla and Denise are 5 years apart.
Betsy and Denise are 6 years apart.
If Alicia is older than Betsy, who is the oldest?

PRACTICE | Answer each question below.

107. ★ In RJ's hockey league, each of the 5 teams plays each other team exactly *twice* during the season. How many total games will be played this season?

107. _____

108. ★★ Three elefinches share a pile of peanuts. The medium elefinch eats twice as many peanuts as the small elefinch, plus one more. The large elefinch eats twice as many peanuts as the medium elefinch, plus one more. After this, there are no more peanuts in the pile. If the large elefinch ate 63 peanuts, how many peanuts were originally in the pile?

108. _____

109. ★★ At the school store, Phyllis buys two pencils and one eraser for 36 cents. Chris buys one pencil and two erasers for 30 cents. How many cents does it cost to buy one pencil and one eraser?

109. _____

HINTS
For Selected Problems

Below are hints to every problem marked with a ★.
Work on the problems for a while before looking at the hints.
The hint numbers match the problem numbers.

CHAPTER 4
Subtraction   6-39

19. How many of the stickers Ralph used were from the big pad?

20. If you subtract Cammie's number minus Nellie's number, what is the hundreds digit of the result?

29. 2 hundreds and 3 tens is the same as 1 hundred and how many tens?

30. 3 hundreds and 1 ten is the same as 2 hundreds and how many tens?

33. How much greater is Lizzie's number than Alex's?

41. What three numbers are used in all four equations?

56. Can any of the numbers with ones digit 9 be crossed out?

57. Which rows and columns are easiest? Are there numbers that are easy to cross out?

58. Which number can be crossed out first in the bottom row? Which number can be crossed out first in the second row?

59. Which of these columns is easiest? Then, what number can be crossed out in the 3rd row? The 5th row?

67. How many more pages were in Richard's first book than in Jeremy's first book?

85. Which two numbers does Hoppy hop between on the second hop?

86. Which two numbers does Hoppy hop between on the fourth hop?

125. What ones digit do we get when we subtract a number that ends in 5 from 74?

126. What must be the ones digit of 6▢?

133. What are the possibilities for the top-right square? What are the possibiities for the bottom-right square? Which of these must you use?

135. What are the possibilities for the square between 10 and 20? What are the possibiities for the square between 20 and 30? Which of these must you use?

136. Working clockwise from 46, the differences between the squares are 2, 3, 4, and 5. In other words, we add or subtract 2, then 3, then 4, and finally 5. How can these differences be used to get from 46 to 48?

137. What are the possibilities for the square between 7 and 2? What are the possibiities for the square between 2 and 3? Which of these must you use?

138. What are the possibilities for the square between 32 and 50? What are the possibiities for the square between 50 and 9? Which of these must you use?

142. Compare 723−237 and 823−137. In 823−137, how much more or less do we start with? How much more or less do we subtract?

143. Compare 632−466 and 622−476. In 622−476, how much more or less do we start with? How much more or less do we subtract?

148. When Polly gives 3 grapes to Molly, what happens to the number of grapes that Polly has? What about the number of grapes Molly has?

149. What is the smallest pair of two-digit numbers that have a difference of 80? The next-smallest?

156. For each difference, can you add or subtract the same amount from both numbers to get 241 and 82?

187. Taking away the three missing numbers one by one is the same as taking away the sum of the three missing numbers all at once.

188. How many 7's do we add to get from the 30th number to the 33rd number?

189. Can you rewrite this subtraction problem as an addition problem? How does that help?

190. What number must go in the first blank?

191. From the equation given, we can draw the following picture:

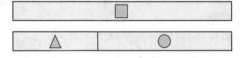

Which of the other statements are described by this picture?

192. What must be the two-digit result?

193. How many inches tall is Stuart?

194. There are six different numbers that can be written using 1, 2, and 3 once each. What are they?

195. What is the smallest number of button presses needed to get the calculator to read 90?

15. Find a good starting place! Which blanks are easy to fill?

16. Find a good starting place! Which blanks are easy to fill?

32. Look at the shaded 8's below. Can you find expressions that include these 8's?

```
      8  +  4  +  8
  18  −  4  −  2  + 10
   −  2  +  1  +  7  −
   6  +  8  + 12  +  5
   +  6  + 11  −  5  +
   6  +  8  −  2  −  7
      4  +  2  +  6
```

33. Look at the shaded 2's below. Can you find expressions that include these 2's?

```
      1  +  1  −  2
  1  +  1  +  1  −  1
  +  1  +  1  +  1  +
  2  −  2  −  2  +  1
  −  2  +  1  −  1  −
  1  +  1  −  2  −  1
      1  +  2  −  1
```

84. What expression gives the team score for Griff and Cliff? For Snark and Clark?

85. How many different ways can you place one pair of parentheses? Two pairs of parentheses? What result does each of these expressions give?

86. Can you use the possibilities you listed on the previous problem?

87. What does $(11+(8-\underline{\hspace{0.5cm}}))$ equal?

88. On which side of the 15 should the equals sign be placed?

112. What expression gives the total number of exercises Erp has already done?

122. How much more is $\mathbf{C}$ than $\mathbf{C}-3$?

127. Does the order that we take $\mathbf{O}$ and 10 away from 20 matter?

128. How much more is $\blacktriangledown+5$ than $\blacktriangledown+2$?

129. How much more is $\square$ than $\square-\blacklozenge$?

133. How does $15+(\blacksquare-\mathbf{C})-(\blacksquare-\mathbf{C})+\blacksquare$ simplify?

154. How must the circles be filled in the top row?

155. How must the circles be filled in the bottom row?

162. How can you write $24-\blacklozenge=\blacklozenge$ as an addition problem?

163. On the left side of the equation, we add $\blacktriangle$ to 5. On the right side of the equation, we add $\blacktriangle$ to $\blacktriangle$. What does this tell us about 5 and $\blacktriangle$?

164. What does $(\mathbf{C}+3)$ equal?

165. Take a guess for the value of $\clubsuit$. Does it work? If not, what might be a good next guess?

167. What are the different ways we can place the $=$ sign?

168. Can you find all the ways to place one pair of parentheses? Try to be organized!

169. There are two ways we can subtract from 5 to get 3: $5-1-1=3$ and $5-2=3$.

170. How much more is $(\mathbf{C}+10)$ than $(\mathbf{C}-10)$?

171. What does $\bigstar$ equal in the equation $\blacktriangle+\bigstar=\blacktriangle+17$?

14. Is the number that fills the blanks more or less than 10?

15. Is the number that fills the blanks more or less than 10?

16. Is the number that fills the blanks more or less than 10?

19. How many total legs are there if Mr. Hayfield has 5 cows?

20. What is the largest 2-digit number? Then, what is the largest number Winnie could have? Then, what is the largest number Grogg could have?

21. Guess and check. Does $\triangle=5$ work?

— *or* —

What is $(\triangle+\triangle+\square)+(\triangle+\square+\square)$? How can this help you find $\triangle+\square$?

22. How many apples are in each bag? Then, how many need to be moved?

33. In the top row, two $\triangle$'s plus one $\square$ equals 50. In the middle row, two $\triangle$'s plus one $\bigcirc$ equals 55. Which is bigger, $\bigcirc$ or $\square$? By how much?

34. In the bottom row, two $\bigcirc$'s plus one $\square$ equals 20. In the left column, two $\bigcirc$'s plus one $\triangle$ equals 22. Which is bigger, $\square$ or $\triangle$? By how much?

40. At the end of each day, how many more acorns does Skippy have than he did the day before?

54. If you get stuck working backwards, try working forwards!

55. Work backwards to find the circle labeled 7. Continuing backwards, there are two circles the knight can reach from there. Which one must be labeled 6?

56. Which circle must be the end of the knight's path?

57. Which circle must be the end of the knight's path?

59. Curt was the last monster to get a sticker. Which monster got a sticker just before Curt?

60. What is the last number that Lizzie wrote down? What is the last number that Lizzie crossed out?

61. b. What is the sum of the digits of the number to the left of 29? What number could this be?

68. Draw a picture. How many feet above the ground is Lenny's home?

69. If we label Amy (A) and Ben (B) five feet apart on the line below, where can we place Kenneth?

71. If Grammy places chocolate chips in 6 cookies first, how can she place pecans in 7 cookies to get the smallest number of cookies with both ingredients?

72. Draw some pictures! How can you make sure you get the most crossings possible among your four roads?

77. How many monsters did Flynn beat that Wynn did not beat?

81. Don't let the number of boxes scare you! Can you use the strategies from the previous problems to solve this one?

85. Draw a picture. Does Gary shake Holden's hand?

91. Can a card that shares two features with only one other card be in the middle of the stack?

96. Can you make 33 cents using just nickels and dimes?

97. How many *incorrect* answers did each student get?

100. Who could be in front of Nick?

102. Draw a picture. How far up the wall is Shelly after each glide?

103. What do you know about addition that might make this problem easier?

104. Draw a picture, then guess and check. How many bridges connect Llama Island to Mama island?

105. Suppose Jorm bought his packs of cards 1 pack at a time. How many cards did Jorm have before he bought the last pack?

106. If we label Carly (C) and Betsy (B) 1 year from each other on the line below, where can we place Alicia?

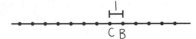

107. How many games would be played if each of the 5 teams played each other team just *once*?

108. How many peanuts did the medium elefinch eat?

109. Guess and check. Can a pencil cost 10 cents?

— *or* —

What is the total cost of everything Phyllis and Chris bought? How can this help you find the cost of one pencil and one eraser?

SOLUTIONS
Chapters 4-6

Take Away 7

1. 3 of the 12 frogs on the lily pad hop into the water. So, we take 3 away from 12. This leaves 9 frogs on the lily pad.

$$12-3=9$$

2. Priscilla pours out 8 of the 15 cups of water in the pitcher. So, we take 8 away from 15. This leaves 7 cups of water in the pitcher.

$$15-8=7$$

3. Tyrone lost 9 of his 17 toy trucks. So, we take 9 away from 17. This leaves 8 toy trucks.

$$17-9=8$$

Place Value 8-9

4. 7 tens minus 2 tens is 5 tens.
4 ones minus 1 one is 3 ones.

5 tens and 3 ones is 53. So, $74-21=53$.

$$74-21=\underset{\text{tens ones}}{7\ 4}-\underset{\text{tens ones}}{2\ 1}=\underset{\text{tens ones}}{5\ 3}$$

5. 6 tens minus 3 tens is 3 tens.
9 ones minus 4 ones is 5 ones.

3 tens and 5 ones is 35. So, $69-34=35$.

$$69-34=\underset{\text{tens ones}}{6\ 9}-\underset{\text{tens ones}}{3\ 4}=\underset{\text{tens ones}}{3\ 5}$$

6. 2 tens minus 1 ten is 1 ten.
5 ones minus 3 ones is 2 ones.

1 ten and 2 ones is 12. So, $25-13=$ **12**.

7. 7 tens minus 3 tens is 4 tens.
9 ones minus 6 ones is 3 ones.

4 tens and 3 ones is 43. So, $79-36=$ **43**.

8. The number 51 has 0 hundreds.

2 hundreds minus 0 hundreds is 2 hundreds.
6 tens minus 5 tens is 1 ten.
5 ones minus 1 one is 4 ones.

2 hundreds, 1 ten, and 4 ones is 214. So, $265-51=$ **214**.

9. 2 hundreds minus 1 hundred is 1 hundred.
8 tens minus 2 tens is 6 tens.
0 ones minus 0 ones is 0 ones.

1 hundred, 6 tens, and 0 ones is 160. So, $280-120=$ **160**.

10. 7 hundreds minus 3 hundreds is 4 hundreds.
6 tens minus 6 tens is 0 tens.
5 ones minus 2 ones is 3 ones.

4 hundreds, 0 tens, and 3 ones is 403. So, $765-362=$ **403**.

11. 9 hundreds minus 9 hundreds is 0 hundreds.
5 tens minus 2 tens is 3 tens.
7 ones minus 6 ones is 1 one.

0 hundreds, 3 tens, and 1 one is 31. So, $957-926=$ **31**.

12. To go from 46 to 35 we subtract 1 ten and 1 one, which is 11.

$$46-\mathbf{11}=35$$

13. To go from 97 to 63 we subtract 3 tens and 4 ones, which is 34.

$$97-\mathbf{34}=63$$

14. To go from 746 to 243 we subtract 5 hundreds, 0 tens, and 3 ones, which is 503.

$$746-\mathbf{503}=243$$

15. We take away 2 tens to get 1 ten.
Since $\underline{3}-2=1$, we start with 3 tens.

We take away 4 ones to get 3 ones.
Since $\underline{7}-4=3$, we start with 7 ones.

So, we start with 3 tens and 7 ones, which is 37.

$$\mathbf{37}-24=13$$

16. We take away 4 tens to get 4 tens.
Since $\underline{8}-4=4$, we start with 8 tens.

We take away 2 ones to get 6 ones.
Since $\underline{8}-2=6$, we start with 8 ones.

So, we start with 8 tens and 8 ones, which is 88.

$$\mathbf{88}-42=46$$

17. We take away 3 hundreds to get 2 hundreds.
Since $\underline{5}-3=2$, we start with 5 hundreds.

We take away 1 ten to get 1 ten.
Since $\underline{2}-1=1$, we start with 2 tens.

We take away 5 ones to get 0 ones.
Since $\underline{5}-5=0$, we start with 5 ones.

So, we start with 5 hundreds, 2 tens, and 5 ones.

$$\mathbf{525}-315=210$$

18. If Ms. Shloop gives 1 candy to each of her 36 students, then she gives away 36 candies all together.

So, Ms. Shloop has $87-36=$ **51** candies left.

19. 110 of the 232 stickers Ralph used came from the small pad. So, $232-110=122$ stickers came from the big pad.

The big pad started with 675 stickers. After 122 of these stickers were used, $675-122=$ **553** stickers remain.

— *or* —

Before Ralph used any stickers, there were $110+675=785$ stickers on both pads combined.

When Ralph used 232 stickers, all the stickers remaining were on the big pad. So, there were $785-232=$ **553** stickers left on the big pad.

20. Cammie's number has 3 more hundreds, 2 more tens, and 1 more one than Nellie's number.

So, Cammie's number minus Nellie's number has hundreds digit 3, tens digit 2, and ones digit 1.

So, the result is **321**.

We can check with an example. If Nellie's number is 555, then Cammie's hundreds digit is $5+3=8$, her tens digit is $5+2=7$, and her ones digit is $5+1=6$. So, Cammie's number is 876, and Cammie's number minus Nellie's number is $876-555=321$. ✓

SUBTRACTION
Breaking   10-11

21. 3 tens and 3 ones is the same as 2 tens and 13 ones.

2 tens and 13 ones minus 1 ten and 5 ones gives $2-1=1$ ten and $13-5=8$ ones, which is 18.

$$33-15 = \underset{\text{tens}}{\underline{3}}\ \underset{\text{ones}}{\underline{3}} - \underset{\text{tens}}{\underline{1}}\ \underset{\text{ones}}{\underline{5}}$$

$$= \underset{\text{tens}}{\underline{2}}\ \underset{\text{ones}}{\underline{13}} - \underset{\text{tens}}{\underline{1}}\ \underset{\text{ones}}{\underline{5}}$$

$$= \boxed{18}$$

22. 4 tens and 1 one is the same as 3 tens and 11 ones.

3 tens and 11 ones minus 2 tens and 6 ones gives $3-2=1$ ten and $11-6=5$ ones, which is 15.

$$41-26 = \underset{\text{tens}}{\underline{4}}\ \underset{\text{ones}}{\underline{1}} - \underset{\text{tens}}{\underline{2}}\ \underset{\text{ones}}{\underline{6}}$$

$$= \underset{\text{tens}}{\underline{3}}\ \underset{\text{ones}}{\underline{11}} - \underset{\text{tens}}{\underline{2}}\ \underset{\text{ones}}{\underline{6}}$$

$$= \boxed{15}$$

23. 8 tens and 6 ones is the same as 7 tens and 16 ones.

7 tens and 16 ones minus 3 tens and 7 ones gives $7-3=4$ tens and $16-7=9$ ones, which is 49.

$$86-37 = \underset{\text{tens}}{\underline{8}}\ \underset{\text{ones}}{\underline{6}} - \underset{\text{tens}}{\underline{3}}\ \underset{\text{ones}}{\underline{7}}$$

$$= \underset{\text{tens}}{\underline{7}}\ \underset{\text{ones}}{\underline{16}} - \underset{\text{tens}}{\underline{3}}\ \underset{\text{ones}}{\underline{7}}$$

$$= \boxed{49}$$

24. 7 tens and 2 ones is the same as 6 tens and 12 ones.

6 tens and 12 ones minus 4 tens and 8 ones gives $6-4=2$ tens and $12-8=4$ ones, which is 24.

$$72-48 = \underset{\text{tens}}{\underline{7}}\ \underset{\text{ones}}{\underline{2}} - \underset{\text{tens}}{\underline{4}}\ \underset{\text{ones}}{\underline{8}}$$

$$= \underset{\text{tens}}{\underline{6}}\ \underset{\text{ones}}{\underline{12}} - \underset{\text{tens}}{\underline{4}}\ \underset{\text{ones}}{\underline{8}}$$

$$= \boxed{24}$$

25. 5 tens and 1 one is the same as 4 tens and 11 ones.

4 tens and 11 ones minus 1 ten and 3 ones gives $4-1=3$ tens and $11-3=8$ ones, which is 38.

So, $51-13=$ **38**.

26. 8 tens and 2 ones is the same as 7 tens and 12 ones.

7 tens and 12 ones minus 2 tens and 5 ones gives $7-2=5$ tens and $12-5=7$ ones, which is 57.

So, $82-25=$ **57**.

27. 9 tens and 4 ones is the same as 8 tens and 14 ones.

8 tens and 14 ones minus 4 tens and 6 ones gives $8-4=4$ tens and $14-6=8$ ones, which is 48.

So, $94-46=$ **48**.

28. 6 tens and 3 ones is the same as 5 tens and 13 ones.

5 tens and 13 ones minus 3 tens and 8 ones gives $5-3=2$ tens and $13-8=5$ ones, which is 25.

So, $63-38=$ **25**.

29. We can't take 7 tens away from 3 tens. But, we can break one of the hundreds in 235 to make 10 tens.

2 hundreds, 3 tens, and 5 ones is the same as 1 hundred, 13 tens, and 5 ones.

1 hundred, 13 tens, and 5 ones minus 0 hundreds, 7 tens, and 3 ones gives $1-0=1$ hundred, $13-7=6$ tens, and $5-3=2$ ones.

So, $235-73=$ **162**.

30. We can't take 8 tens away from 1 ten. But, we can break one of the hundreds in 317 to make 10 tens.

3 hundreds, 1 ten, and 7 ones is the same as 2 hundreds, 11 tens, and 7 ones.

2 hundreds, 11 tens, and 7 ones minus 1 hundred, 8 tens, and 0 ones gives $2-1=1$ hundred, $11-8=3$ tens, and $7-0=7$ ones.

So, $317-180=$ **137**.

31. Polly eats 26 of his 54 crackers. So, he has $54-26$ crackers left.

5 tens and 4 ones is the same as 4 tens and 14 ones.

4 tens and 14 ones minus 2 tens and 6 ones gives $4-2=2$ tens and $14-6=8$ ones, which is 28.

So, Polly has $54-26=$ **28** crackers left.

32. If we take away the 72 students who left to play, the remaining students are the ones who did not play. So, $215-72$ of Ms. Melody's students did not play.

To subtract, we break one of the hundreds in 215 to make 10 tens.

2 hundreds, 1 ten, and 5 ones is the same as 1 hundred, 11 tens, and 5 ones.

1 hundred, 11 tens, and 5 ones minus 0 hundreds, 7 tens, and 2 ones gives $1-0=1$ hundred, $11-7=4$ tens, and $5-2=3$ ones.

So, $215-72=$ **143** of Ms. Melody's students did not play.

33. Lizzie's number has 2 more tens and 2 fewer ones than Alex's number.

So, to get from Alex's number to Lizzie's number, we can add 2 tens, or 20, and take away 2 ones.

Adding 20 then taking away 2 is the same as adding 18.

So, Lizzie's number is 18 more than Alex's number.

This means that taking away Alex's number from Lizzie's number leaves **18**.

We can check with an example. If Alex's number is 55, then Lizzie's tens digit is $5+2=7$, and her ones digit is $5-2=3$. So, Lizzie's number is 73, and Lizzie's number minus Alex's number is $73-55=18$. ✓

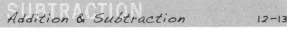

Addition & Subtraction 12–13

34. The 13 coins that landed heads plus the coins that landed tails give 21 coins all together.

$$13 + \boxed{} = 21$$

Since $13+\underline{8}=21$, Stu flipped **8** coins that landed tails.

— *or* —

If we take away the 13 coins that landed heads, only the coins that landed tails will be left.

$$21 - 13 = \boxed{}$$

Since $21-13=\underline{8}$, Stu flipped **8** coins that landed tails.

35. The 30 minutes spent jogging plus the minutes spent walking equal the 55 minutes spent walking and jogging.

$$30 + \boxed{} = 55$$

Since $30+\underline{25}=55$, Mary spent **25** minutes walking.

— *or* —

If we take away the 30 minutes Mary spent jogging, only the minutes that she spent walking will be left.

$$55 - 30 = \boxed{}$$

Since $55-30=\underline{25}$, Mary spent **25** minutes walking.

36. Brett has two 20-dollar bills. So, Brett has $20+20=40$ dollars. Brett's 40 dollars plus Chuck's dollars equals the 83 dollars they have together.

$$40 + \boxed{} = 83$$

Since $40+\underline{43}=83$, Chuck has **43** dollars.

— *or* —

If we take away Brett's 40 dollars from the 83 total dollars, only Chuck's dollars will be left.

$$83 - 40 = \boxed{}$$

Since $80-40=\underline{43}$, Chuck has **43** dollars.

37. The length of the 14-foot piece plus the length of the other piece equals the length of the original 32-foot rope.

$$14 + \boxed{} = 32$$

Since $14+\underline{18}=32$, the other piece of rope is **18** feet long.

— *or* —

If we take away the 14-foot piece from the 32-foot rope, only the length of the other piece will be left.

$$32 - 14 = \boxed{}$$

Since $32-14=\underline{18}$, the other piece of rope is **18** feet long.

38. We complete the four statements as shown below.

$$\boxed{27} + \boxed{56} = \boxed{\textbf{83}}$$
$$\boxed{56} + \boxed{\textbf{27}} = \boxed{83}$$
$$\boxed{\textbf{83}} - \boxed{56} = \boxed{27}$$
$$\boxed{83} - \boxed{\textbf{27}} = \boxed{56}$$

39. We complete the four statements as shown below.

$$\boxed{88} + \boxed{44} = \boxed{\textbf{132}}$$
$$\boxed{44} + \boxed{\textbf{88}} = \boxed{132}$$
$$\boxed{132} - \boxed{\textbf{44}} = \boxed{88}$$
$$\boxed{\textbf{132}} - \boxed{88} = \boxed{\textbf{44}}$$

40. The third statement is easiest to complete. So, we start there: $\boxed{123} + \boxed{58} = \boxed{\textbf{181}}$. Then, we complete the remaining three statements as shown below.

$$\boxed{\textbf{181}} - \boxed{58} = \boxed{123}$$
$$\boxed{\textbf{181}} - \boxed{123} = \boxed{58}$$
$$\boxed{123} + \boxed{58} = \boxed{\textbf{181}}$$
$$\boxed{58} + \boxed{\textbf{123}} = \boxed{\textbf{181}}$$

41. The three numbers to be used are 25, 39, and 64. We use these three numbers to make four different statements as shown below.

$$\boxed{39} + \boxed{\textbf{25}} = \boxed{\textbf{64}}$$
$$\boxed{\textbf{25}} + \boxed{39} = \boxed{64}$$
$$\boxed{64} - \boxed{\textbf{25}} = \boxed{\textbf{39}}$$
$$\boxed{64} - \boxed{\textbf{39}} = \boxed{\textbf{25}}$$

Three-Four 14–15

42. **43.**

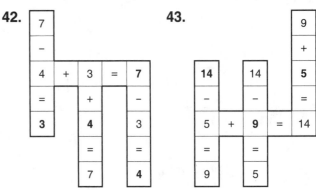

44.

28 − 12 = 16
−
16 + 12 = 28
=
12 + 16 = 28

45.

24   87
+   −
63 + 24 = 87
=   [ ]   =
87 − 63 = 24

46.

56
+
18   74
=   −
74 = 18 + 56
        =
74 − 56 = 18

47.

222
+
111   111
=   +
333 − 222 = 111
         =
333 − 111 = 222

48.

562   237
−   +
237   562 − 325 = 237
=   =
325 + 237 = 562

49.

238 + 686 = 924
        +   [ ]   −
        238   686
        =   [ ]   =
924 − 238 = 686

50. In the top row, 5 and 5 are the only numbers we can add to get a sum of 10. So, we cross out the 2's. We also circle the 5's to remind ourselves that these numbers must be used. We do the same thing in the right column.

In the bottom row, one 2 is crossed out. The remaining numbers are 4, 4, and 2. Since $4+4+2=10$, we circle all three of these numbers. We do the same thing in the left column.

In the third row, one 4 is circled. So, the other numbers we circle must sum to $10-4=6$. Only two 3's remain. Since $3+3=6$, we circle both 3's. We do the same thing in the second column.

We complete the puzzle by crossing out the remaining 3.

We check to make sure that the circled numbers in every row and column sum to 10.

51. In the bottom row, all four numbers sum to $7+6+5+4=22$. So, we circle all four numbers.

Similarly, in the third column, all four numbers sum to $3+4+10+5=22$. So, we circle all four numbers.

In the left column, 7 is circled. So, the other numbers we circle sum to $22-7=15$. We cannot get a sum of 15 using the 5 or 3. So, we circle 15 and cross out the 5 and 3.

In the second column, 6 is circled. So, the other numbers we circle sum to $22-6=16$. Only 4 and 12 sum to 16. So, we circle 4 and 12 and cross out 11.

In the right column, 4 is circled. So, the other numbers we circle sum to $22-4=18$. We cannot get a sum of 18 using 9 or 1. So, we circle 18, and cross out 9 and 1.

The completed puzzle is shown to the right.

We use the strategies discussed in the previous problems to complete the puzzles below.

52.

22	14	22	44
22	32	34	44
22	34	32	44
22	22	64	44

53.

4	106	2	103
101	6	102	5
7	108	1	104
107	8	109	3

54. We notice that the grid contains only three different numbers: 100, 20, and 3. Since $100+20+3$ gives the target sum 123, we must use one 100, one 20, and one 3 in each row and column.

In the left column, we have one 100, one 20, and two 3's. We don't know which 3 to circle yet, but we can circle 100 and 20.

100	20	100	3
3	20	20	100
20	3	100	3
3	100	3	20

Similarly, we can circle numbers in the other three columns as shown.

100	20	100	3
3	20	20	100
20	3	100	3
3	100	3	20

Then, we use the same strategy to circle numbers in the rows as shown.

100	20	100	3
3	20	20	100
20	3	100	3
3	100	3	20

Finally, we cross out the unused numbers to complete the puzzle as shown.

100	20	100	3
3	20	20	100
20	3	100	3
3	100	3	20

55. We use the strategies discussed in the previous problems to complete the puzzle as shown to the right.

4	1	3	3
4	3	4	3
3	4	3	4
3	3	4	1

56. Each row has four numbers with ones digit 0 and one number with ones digit 9. So, to get a sum of 99 in each row, we must use each number that has ones digit 9.

50	49	20	20	50
10	10	20	20	39
10	10	39	30	30
39	40	40	30	30
50	40	40	19	50

The remaining numbers in the second row sum to $99-39=60$. Since $10+10+20+20=60$, we circle all of the numbers in this row.

The remaining numbers in the third row sum to $99-39=60$. Only 30 and 30 sum to 60, so we circle the two 30's and cross out the two 10's.

50	49	20	20	50
10	10	20	20	39
10	10	39	30	30
39	40	40	30	30
50	40	40	19	50

The remaining numbers in the fourth row sum to $99-39=60$. Only 30 and 30 sum to 60, so we circle the two 30's and cross out the two 40's.

The remaining numbers in the fifth row sum to $99-19=80$. Only 40 and 40 sum to 80, so we circle the two 40's and cross out the two 50's.

50	49	20	20	50
10	10	20	20	39
10	10	39	30	30
39	40	40	30	30
50	40	40	19	50

In the left column, we circle 50 to get a sum of $50+10+39=99$.

In the top row, 50 and 49 are circled, and $50+49=99$. So, we cross out the three remaining numbers in this row.

50	49	20	20	50
10	10	20	20	39
10	10	39	30	30
39	40	40	30	30
50	40	40	19	50

The completed puzzle is shown to the right.

50	49	20	20	50
10	10	20	20	39
10	10	39	30	30
39	40	40	30	30
50	40	40	19	50

57. We consider the large number in the second column: 26. If we circle 26, then the other numbers we circle must sum to $28-26=2$. There are no numbers in this column that sum to 2. So, we cross out 26. Then, 20 and 8 are the only numbers that sum to 28. So, we circle 20 and 8, and cross out 9 and 6.

10	9	9	9	19
6	26	11	17	9
6	8	23	5	9
6	20	2	14	9
16	9	6	6	10

We consider the large number in the middle column: 23. If we circle 23, then the other numbers we circle must sum to $28-23=5$. There are no numbers in this column that sum to 5. So, we cross out 23. Then, the remaining numbers sum to $9+11+2+6=28$. So, we circle these numbers.

10	9	9	9	19
6	26	11	17	9
6	8	23	5	9
6	20	2	14	9
16	6	6	6	10

Then, there is only one way to circle and cross out numbers in the bottom four rows so that the sum of the circled numbers in each row is 28.

Finally, we complete the first, fourth, and fifth columns to finish the puzzle as shown.

58. In the left column, if we circle 1, then the other numbers we circle sum to $8-1=7$. But, we cannot get a sum of 7 using 2's and 4's. So, we cross out 1.

In the second row, if we circle 2, then the other numbers we circle sum to $8-2=6$. But, we cannot get a sum of 6 using 4's and 8's. So, we cross out 2.

We do the same thing in the fourth column.

In the left column, the remaining numbers are 2, 2, and 4. These numbers sum to 8. So, we circle 2, 2, and 4.

We do the same thing in the bottom row.

In the top row, one 4 is circled. So, the other numbers we circle sum to $8-4=4$. So, we circle the other 4 and cross out the three 8's.

We do the same thing in the right column.

In the middle row, one 2 is circled. So, the other numbers we circle sum to $8-2=6$. Only 2 and 4 sum to 6. So, we circle 2 and one of the 4's. We don't know which 4 to circle yet, but we can circle the 2.

We do the same thing in the middle column.

The circled numbers in the fourth row sum to $2+2+4=8$. So, we cross out the remaining numbers in this row.

We do the same thing in the second column.

In the second row, every number is crossed out except for two 4's. So, we circle these two 4's.

We do the same thing in the fourth column.

Finally, we cross out the 4 in the middle of the grid to complete the puzzle as shown

59. In the right column, all five 5's sum to $5+5+5+5+5=25$. So, we circle all of the 5's.

In the middle row, one 5 is circled. So, the other numbers we circle sum to $25-5=20$. The remaining numbers are 5, 10, 10, and 10. We cannot get a sum of 20 if we circle the other 5. So, we cross out this 5.

We do the same thing in the bottom row.

In the left column, one 5 is crossed out, and the remaining numbers are 5, 5, 10, and 10. We can only get a sum of 25 by circling two 10's and one 5. We don't know which 5 to circle yet, but we can circle both 10's.

In the fourth column, one 5 is crossed out, and the remaining numbers are 5, 10, 10, and 10. We can only get a sum of 25 by circling two 10's and one 5. We don't know which two 10's to circle yet, but we can circle the remaining 5.

In the top row, one 10 and two 5's are circled, giving a sum of 10+5+5 = 20. So, the other numbers we circle sum to 25−20 = 5. So, we circle the remaining 5 and cross out the remaining 10.

10	X	5	5	5
5	5	5	10	5
5	10	10	10	5
5	5	10	10	5
10	10	10	5	5

We continue using these strategies to complete the puzzle, as shown to the right.

10	X	5	5	5
5	5	X	10	5
X	10	10	X	5
X	X	10	10	5
10	10	X	X	5

SUBTRACTION
Difference
18-19

60. We are looking for the number that can be added to 23 to give us 65.

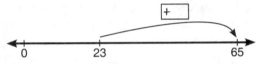

We can find the missing number in this addition problem using subtraction: 65−23 = <u>42</u>.

So, 65 is greater than 23 by **42**.

61. To find the difference between two numbers, we subtract.

555−234 = 321. So, the difference between 555 and 234 is **321**.

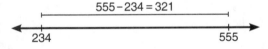

62. The difference between two numbers tells us how far apart they are on the number line. So, the two numbers with the greatest difference are the two numbers that are farthest apart on the number line.

Among the numbers given, 19 is the smallest and 156 is the largest, so they are farthest apart on the number line.

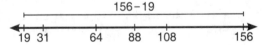

So, 19 and 156 are the two numbers with the greatest difference.

64   (19)   108   31   (156)   88

63. To make the biggest difference, we want the 3-digit number to be as large as possible, and the 2-digit number to be as small as possible.

The largest 3-digit number is 999.

The smallest 2-digit number is 10.

So, 999−10 = **989** is the greatest possible difference between a 3-digit number and a 2-digit number.

64. Jill fetched 43−26 = **17** more pails of water than Jack.

65. Humpback Hill is 460−350 = **110** feet taller than Knobby Knoll.

66. Shelly is 7 years older than Tamika. Since Rebecca is 9 years older than Shelly, Rebecca is 7+9 = 16 years older than Tamika.

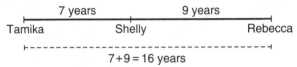

So, the difference between Rebecca's age and Tamika's age is **16** years.

67. All together, Jeremy read 222+328 = 550 pages. All together, Richard read 236+348 = 584 pages.

So, Richard read 584−550 = **34** more pages than Jeremy.

— *or* —

Richard's first book is 236−222 = 14 more pages than Jeremy's first book.

Richard's second book is 348−328 = 20 more pages than Jeremy's second book.

So, Richard read 14+20 = **34** more pages than Jeremy all together.

SUBTRACTION
Counting Up
20-21

68. 100 is 1 more than 99.
107 is 7 more than 100.

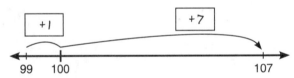

So, 107 is 1+7 = 8 more than 99.
This means that 107−99 = **8**.

69. 500 is 7 more than 493.
502 is 2 more than 500.

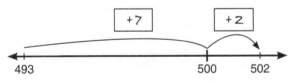

So, 502 is 7+2 = 9 more than 493.
This means that 502−493 = **9**.

70. 100 is 25 more than 75.
111 is 11 more than 100.

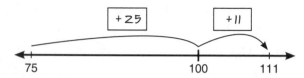

So, 111 is 25+11 = 36 more than 75.
This means that 111−75 = **36**.

71. 200 is 50 more than 150.
241 is 41 more than 200.

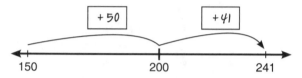

So, 241 is 50+41 = 91 more than 150.
This means that 241−150 = **91**.

72. 100 is 8 more than 92.
200 is 100 more than 100.
203 is 3 more than 200.

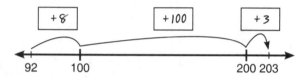

So, 203 is 8+100+3 = 111 more than 92.
This means that 203−92 = **111**.

73. 500 is 5 more than 495.
800 is 300 more than 500.
804 is 4 more than 800.

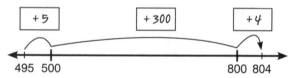

So, 804 is 5+300+4 = 309 more than 495.
This means that 804−495 = **309**.

74. We count up by 3 to get from 97 to 100.
We count up by 25 to get from 100 to 125.

All together, we count up by 3+25 = 28 to get from 97 to 125. So, 125−97 = **28**.

75. We count up by 15 to get from 185 to 200.
We count up by 40 to get from 200 to 240.

All together, we count up by 15+40 = 55 to get from 185 to 240. So, 240−185 = **55**.

76. We count up by 10 to get from 290 to 300.
We count up by 400 to get from 300 to 700.

All together, we count up by 10+400 = 410 to get from 290 to 700. So, 700−290 = **410**.

77. We count up by 25 to get from 175 to 200.
We count up by 31 to get from 200 to 231.

All together, we count up by 25+31 = 56 to get from 175 to 231. So, 231−175 = **56**.

78. We count up by 50 to get from 350 to 400.
We count up by 100 to get from 400 to 500.
We count up by 43 to get from 500 to 543.

All together, we count up by 50+100+43 = 193 to get from 350 to 543. So, 543−350 = **193**.

79. We count up by 9 to get from 591 to 600.
We count up by 300 to get from 600 to 900.
We count up by 9 to get from 900 to 909.

All together, we count up by 9+300+9 = 318 to get from 591 to 909. So, 909−591 = **318**.

80. The first hop is a distance of 41. Only 29 and 70 are 41 apart: 70−29 = 41. So, we either start at 70 and hop to 29, or start at 29 and hop to 70. We draw this hop on the number line as shown below.

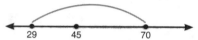

The next distance is 25. Only 45 and 70 are 25 apart: 70−45 = 25. So, Hoppy starts at 29, hops to 70, then hops to 45.

We draw the complete path as shown below.

81. The first distance is 26. Only 53 and 79 are 26 apart: 79−53 = 26. So, the first hop is either from 53 to 79, or from 79 to 53.

The next distance is 53. Only 26 and 79 are 53 apart: 79−26 = 53. So, Hoppy starts at 53, hops to 79, then hops to 26.

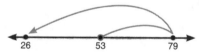

82. The first distance is 43. Only 42 and 85 are 43 apart: 85−42 = 43. So, the first hop is between 42 and 85.

The next distance is 79. Only 85 and 6 are 79 apart: 85−6 = 79. So, the next hop is from 85 to 6.

The last distance is 60, and 6 and 66 are 60 apart: 66−6 = 60. So, the last hop is from 6 to 66.

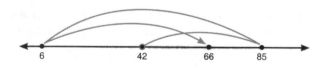

83. The first distance is 10. There are two pairs of numbers that are 10 apart: (15 and 25) and (25 and 35).

So, we start by looking at the second distance, 71. Only 25 and 96 are 71 apart: $96 - 25 = 71$. So, we know there is a hop between 25 and 96.

The third distance is 81. Only 96 and 15 are 81 apart: $96 - 15 = 81$. So, Hoppy hops from 25 to 96, then from 96 to 15.

Since Hoppy hops to every number once, the first hop must be from 35 to 25, which correctly gives a distance of 10.

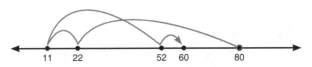

84. The first distance is 58. Only 22 and 80 are 58 apart: $80 - 22 = 58$. So, the first hop is between 22 and 80.

The next distance is 11. Only 22 and 11 are 11 apart: $22 - 11 = 11$. So, the next hop is from 22 to 11.

The next distance is 41. Only 11 and 52 are 41 apart: $52 - 11 = 41$. So, the next hop is from 11 to 52.

The last distance is 8, and 52 and 60 are 8 apart: $60 - 52 = 8$. So, the last hop is from 52 to 60.

85. The first distance is 35. There are two pairs of numbers that are 35 apart: (18 and 53) and (53 and 88).

So, we start by looking at the second distance, 37. Only 51 and 88 are 37 apart: $88 - 51 = 37$. So, we know there is a hop between 51 and 88.

The third distance is 33. Only 51 and 18 are 33 apart: $51 - 18 = 33$. So, we know there is a hop from 51 to 18.

The last distance is 39. Only 18 and 57 are 39 apart: $57 - 18 = 39$. So, we know the last hop is from 18 to 57.

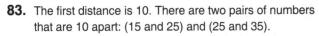

Since Hoppy hops to every number once, we know the first hop is from 53 to 88, which correctly gives a distance of 35.

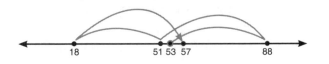

86. The first and third distances are 39. There are two pairs of numbers that are 39 apart: (20 and 59) and (60 and 99).

The second distance is 40. There are two pairs of numbers that are 40 apart: (20 and 60) and (59 and 99).

So, we start by looking at the last distance, 12. Only 59 and 71 are 12 apart: $71 - 59 = 12$. So, the last hop is between 59 and 71.

Working backwards, we look at the third distance, 39. Of 59 and 71, only 59 is 39 apart from another number, 20. So, the third hop is from 20 to 59.

We keep working backwards. The second distance is 40. Since $60 - 20 = 40$, the second hop is from 60 to 20.

Finally, since $99 - 60 = 39$, the first hop is from 99 to 60.

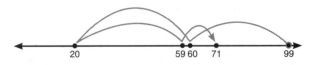

Subtract then Add   24-25

87. After Brian gives Ted 100 dollars, Brian has $350 - 100$ dollars. Then, after Brian takes 5 dollars back, Brian has **$350 - 100 + 5$** dollars.

Giving away 100 dollars, then taking 5 dollars back is the same as giving away 95 dollars. So, Brian has **$350 - 95$** dollars.

($350 - 95$) $350 - 100 - 5$ ($350 - 100 + 5$) $350 - 105$

88. After 100 students leave to go to lunch, there are $284 - 100$ students in the gym. Then, after 12 students return, there are **$284 - 100 + 12$** students still in the gym.

100 students leaving, then 12 students returning is the same as 88 students leaving. So, there are **$284 - 88$** students still in the gym.

$284 - 112$ ($284 - 88$) $284 - 100 - 12$ ($284 - 100 + 12$)

89. Taking away 198 is the same as taking away 200, then giving 2 back. So, $755 - 198$ is the same as $755 - 200 + 2$.

$755 + 200 + 2$ ($755 - 200 + 2$) $755 - 200 - 2$ $755 + 200 - 2$

90. Subtracting 96 is the same as subtracting 100, then adding **4**.

91. Subtracting 190 is the same as subtracting **200**, then adding 10.

92. Subtracting **48** is the same as subtracting 50, then adding 2.

93. $261-94=\boxed{261}-\boxed{100}+\boxed{6}=\boxed{167}$.

94. $72-39=\boxed{72}-\boxed{40}+\boxed{1}=\boxed{33}$.

95. $450-195=\boxed{450}-\boxed{200}+\boxed{5}=\boxed{255}$.

96. Subtracting 97 is the same as subtracting 100, then adding 3. So,
$$123-97=123-100+3$$
$$=23+3$$
$$=\textbf{26}.$$

97. Subtracting 49 is the same as subtracting 50, then adding 1. So,
$$383-49=383-50+1$$
$$=333+1$$
$$=\textbf{334}.$$

98. Subtracting 195 is the same as subtracting 200, then adding 5. So,
$$650-195=650-200+5$$
$$=450+5$$
$$=\textbf{455}.$$

99. Subtracting 75 is the same as subtracting 100, then adding 25. So,
$$333-75=333-100+25$$
$$=233+25$$
$$=\textbf{258}.$$

100. Subtracting 299 is the same as subtracting 300, then adding 1. So,
$$992-299=992-300+1$$
$$=692+1$$
$$=\textbf{693}.$$

101. Subtracting 390 is the same as subtracting 400, then adding 10. So,
$$868-390=868-400+10$$
$$=468+10$$
$$=\textbf{478}.$$

SUBTRACTION
Review 26

We show one strategy for solving each subtraction problem below. You may have used a different strategy to arrive at the same answer.

102. Subtracting 96 is the same as subtracting 100, then adding 4. So, $355-96=355-100+4=\textbf{259}$.

103. We subtract by place value.

3 hundreds minus 1 hundred is 2 hundreds.
5 tens minus 2 tens is 3 tens.
5 ones minus 2 ones is 3 ones.

2 hundreds, 3 tens, and 3 ones is 233. So, $355-122=\textbf{233}$.

104. We count up by 25 to get from 275 to 300.
We count up by 55 to get from 300 to 355.

All together, we count up by $25+55=80$ to get from 275 to 355. So, $355-275=\textbf{80}$.

105. We count up by 14 to get from 486 to 500.
We count up by 17 to get from 500 to 517.

All together, we count up by $14+17=31$ to get from 486 to 517. So, $517-486=\textbf{31}$.

106. Subtracting 198 is the same as subtracting 200, then adding 2. So, $517-198=517-200+2=\textbf{319}$.

107. We subtract by place value.

5 hundreds minus 2 hundreds is 3 hundreds.
1 ten minus 1 ten is 0 tens.
7 ones minus 4 ones is 3 ones.

3 hundreds, 0 tens, and 3 ones is 303. So, $517-214=\textbf{303}$.

108. We subtract by place value.

8 hundreds minus 7 hundreds is 1 hundred.
7 tens minus 6 tens is 1 ten.
6 ones minus 5 ones is 1 one.

1 hundred, 1 ten, and 1 one is 111. So, $876-765=\textbf{111}$.

109. We count up by 20 to get from 780 to 800.
We count up by 76 to get from 800 to 876.

All together, we count up by $20+76=96$ to get from 780 to 876. So, $876-780=\textbf{96}$.

110. Subtracting 90 is the same as subtracting 100, then adding 10. So, $876-90=876-100+10=\textbf{786}$.

111. We count up by 30 to get from 470 to 500.
We count up by 39 to get from 500 to 539.

All together, we count up by $30+39=69$ to get from 470 to 539. So, $539-470=\textbf{69}$.

112. Subtracting 295 is the same as subtracting 300, then adding 5. So, $539-295=539-300+5=\textbf{244}$.

113. We subtract by place value.

5 hundreds minus 2 hundreds is 3 hundreds.
3 tens minus 2 tens is 1 ten.
9 ones minus 6 ones is 3 ones.

3 hundreds, 1 ten, and 3 ones is 313. So, $539-226=\textbf{313}$.

114. We subtract by place value.

7 hundreds minus 6 hundreds is 1 hundred.
0 tens minus 0 tens is 0 tens.
7 ones minus 4 ones is 3 ones.

1 hundred, 0 tens, and 3 ones is 103. So, $707-604=\textbf{103}$.

115. We count up by 41 to get from 659 to 700.
We count up by 7 to get from 700 to 707.

All together, we count up by $41+7=48$ to get from 659 to 707. So, $707-659=\textbf{48}$.

116. Subtracting 280 is the same as subtracting 300, then adding 20. So, $707-280=707-300+20=\textbf{427}$.

117. Since $19 - 7 = 12$, we have

$$19 - \boxed{7} = 12.$$

118. Since $25 - 13 = 12$, we have

$$25 - 1\boxed{3} = 12.$$

119. Since $47 - 33 = 14$, we have

$$47 - \boxed{3}3 = 14.$$

120. In the ones place, we have 7 ones − ☐ ones = 5 ones. Since $7 - \boxed{2} = 5$, we fill the ones blank with 2.

This gives $\boxed{}7 - 3\boxed{2} = 55$. Since $87 - 32 = 55$, we have

$$\boxed{8}7 - 3\boxed{2} = 55.$$

121. In the ones place, we have ☐ ones − 6 ones = 2 ones. Since $\boxed{8} - 6 = 2$, we fill the ones blank with 8.

This gives $9\boxed{8} - \boxed{}6 = 12$. Since $98 - 86 = 12$, we have

$$9\boxed{8} - \boxed{8}6 = 12.$$

122. In the ones place, we have ☐ ones − 4 ones = 5 ones. Since $\boxed{9} - 4 = 5$, we fill the ones blank with 9.

This gives $8\boxed{9} - 34 = \boxed{}5$. Since $89 - 34 = 55$, we have

$$8\boxed{9} - 34 = \boxed{5}5.$$

123. When we subtract ☐ from 34, the result must be between $34 - 9 = 25$ and $34 - 0 = 34$. The only number between 25 and 34 that ends in 8 is 28.

So, we have $34 - \boxed{} = \boxed{2}8$.
Then, since $34 - 6 = 28$, we have

$$34 - \boxed{6} = \boxed{2}8.$$

124. When we subtract 8 from a number that ends in 1, we always get a result that ends in 3. For example, $61 - 8 = 53$, and $131 - 8 = 123$.

This is because we must break a ten into 10 ones before subtracting, which gives us $11 - 8 = 3$ in the ones place.

So, we have $\boxed{}1 - 8 = 4\boxed{3}$.
Then, since $51 - 8 = 43$, we have

$$\boxed{5}1 - 8 = 4\boxed{3}.$$

125. When we subtract a number that ends in 5 from 74, we always get a result that ends in 9. For example, $7\underline{4} - 3\underline{5} = 3\underline{9}$, and $7\underline{4} - 1\underline{5} = 5\underline{9}$.

This is because we must break a ten into 10 ones before subtracting, which gives us $14 - 5 = 9$ in the ones place.

So, we have $74 - \boxed{}5 = 1\boxed{9}$.
Then, since $74 - 55 = 19$, we have

$$74 - \boxed{5}5 = 1\boxed{9}.$$

126. In the ones place, we have ☐ ones − 7 ones = 4 ones. There is no digit we can take 7 away from to get 4. But, we know $\underline{11} - 7 = 4$.

So, to subtract $6☐ - ☐7$, we break a ten in $6☐$ to make 5 tens and $\underline{11}$ ones, which is $6\boxed{1}$.

So, we have $6\boxed{1} - ☐7 = 34$.
Then, since $61 - 27 = 34$, we have

$$6\boxed{1} - \boxed{2}7 = 34.$$

— *or* —

If $6☐ - ☐7 = 34$, then $34 + ☐7 = 6☐$.

The only number ☐7 that we can add to 34 to get a result in the 60's is 27. So, we have $34 + \boxed{2}7 = 6☐$.

Since $34 + \boxed{2}7 = 6\boxed{1}$, we have

$$6\boxed{1} - \boxed{2}7 = 34.$$

127. $34 - 9 = 25$. So, we fill the empty triangle with 25.

128. $85 - 29 = 56$. So, we fill the empty triangle with 56.

129. $72 - 18 = 54$. So, we fill the empty triangle with 54.

Then, we consider the empty square.

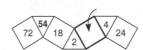

The difference of the empty square and 18 is 2. This means that the empty square is either 2 more or 2 less than 18. So, the empty square is either $18 + 2 = 20$ or $18 - 2 = 16$.

The difference of the empty square and 24 is 4. So, the empty square is either $24 + 4 = 28$ or $24 - 4 = 20$.

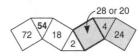

Only 20 correctly gives both differences. So, the empty square is 20.

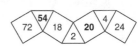

130. We use the strategies from the previous problem to complete the puzzle as shown below.

131. 33 – 19 = 14, so we fill the top empty triangle with 14.
33 – 14 = 19, so we fill the left empty triangle with 19.

The difference of the empty square and 19 is 11.
So, the empty square is either 19 + 11 = 30 or 19 – 11 = 8.

The difference of the empty square and 14 is 6.
So, the empty square is either 14 + 6 = 20 or 14 – 6 = 8.

Only 8 correctly gives both differences. So, the empty square is 8.

132. The difference of the bottom empty square and 58 is 3.
So, the bottom empty square is 58 + 3 = 61 or 58 – 3 = 55.

The difference of the bottom empty square and 50 is 5.
So, the bottom empty square is 50 + 5 = 55 or 50 – 5 = 45.

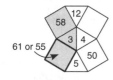 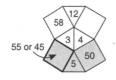

Only 55 correctly gives both differences. So, the bottom empty square is 55.

Then, we use the same strategy to finish the puzzle as shown below.

133. 37 – 25 = 12, so we fill the empty triangle with 12.

The difference of the top empty square and 25 is 15.
So, the top empty square is 25 + 15 = 40 or 25 – 15 = 10.

The difference of the bottom empty square and 37 is 13.
So, the bottom empty square is 37 + 13 = 50 or 37 – 13 = 24.

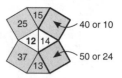

The difference of the two empty squares is 14. Among our possibilities, only 10 and 24 have a difference of 14. So, the top square is 10 and the bottom square is 24.

134. We use the strategies discussed in previous problems to complete the puzzle as shown.

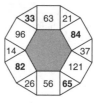

135. We consider the empty squares on the top and the right.

The difference of the top empty square and 15 is 10.
So, the top empty square is 15 + 10 = 25 or 15 – 10 = 5.

The difference of the right empty square and 75 is 30.
So, the right empty square is 75 + 30 = 105 or 75 – 30 = 45.

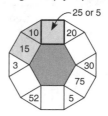

The difference of these two empty squares is 20. Among our possibilities, only 25 and 45 have a difference of 20. So, the top square is 25 and the right square is 45.

We use the same strategy to fill the remaining two empty squares, as shown below.

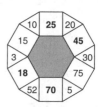

136. We consider the left empty square.

The difference of the left empty square and 46 is 33.
So, the left empty square is 46 + 33 = 79 or 46 – 33 = 13.

The difference of the left empty square and 48 is 31.

So, the left empty square is 48+31 = 79 or 48−31 = 17.

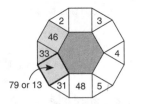 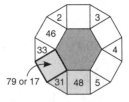

Only 79 correctly gives both differences. So, the left empty square is 79.

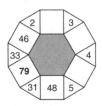

Then, we look at the square filled with 46. The difference between this square and the empty square to its right is 2. So, starting with 46, we add 2 or subtract 2 to get the empty square to its right.

Similarly, moving clockwise around the figure:
We add or subtract 3 to get the next empty square.
We add or subtract 4 to get the next empty square.
We add or subtract 5 to get the next square, which is 48.

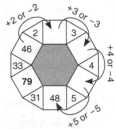

So, starting with 46, we add or subtract 2, 3, 4, and 5 to end up with 48, which is 2 more than 46.

We can only get an total increase of 2 by adding 3 and 5 and subtracting 2 and 4. So, we complete the puzzle as shown.

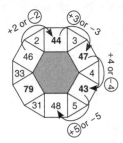

137. We consider the two empty squares on the bottom.

The difference of the left empty square and 20 is 7. So, the left empty square is 20+7 = 27 or 20−7 = 13.

The difference of the right empty square and 18 is 3. So, the right empty square is 18+3 = 21 or 18−3 = 15.

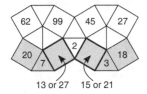

The difference of the two empty squares is 2. Among our possibilities, only 13 and 15 have a difference of 2. So, the left square is 13 and the right square is 15.

We then fill the empty triangles to complete the puzzle as shown.

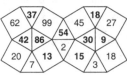

138. We consider the two bottom empty squares.

The difference of the bottom-left empty square and 43 is 32. So, the bottom-left empty square is 43+32 = 75 or 43−32 = 11.

The difference of the bottom-right empty square and 34 is 9. So, the bottom-right empty square is 34+9 = 43 or 34−9 = 25.

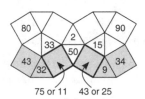

The difference between the bottom empty squares is 50. Among our possibilities, only 75 and 25 have a difference of 50. So, the bottom-left empty square is 75 and the bottom-right empty square is 25.

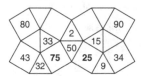

The difference of the top-left empty square and 75 is 33. So, the top-left empty square is 75+33 = 108 or 75−33 = 42.

The difference of the top-right empty square and 25 is 15. So, the top-right empty square is 25+15 = 40 or 25−15 = 10.

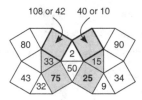

The difference between the top empty squares is 2. Among our possibilities, only 42 and 40 have a difference of 2. So, the top-left empty square is 42 and the top-right empty square is 40.

Finally, we fill the empty triangles to complete the puzzle as shown.

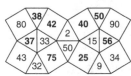

139. We use 613−548 to find 614−548.

In 614−548, we start with 1 more than in 613−548, but we subtract the same amount.

Starting with 1 more makes our result greater by 1.

613−548 = 65. So, we have
614−548 = 65+1 = **66**.

— *or* —

We consider the difference on the number line.

614 is 1 unit farther from 548 than 613 is from 548. So, 614−548 is 1 more than 613−548.

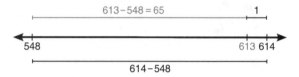

613−548 = 65. So, we have 614−548 = 65+1 = **66**.

140. We use 921−267 to find 921−367.

In 921−367, we start with the same number as in 921−267, but we subtract 100 more.

Subtracting 100 more makes our result smaller by 100.

921−267 = 654. So, we have
921−367 = 654−100 = **554**.

— *or* —

We consider the difference on the number line.

367 is 100 units closer to 921 than 267 is to 921. So, 921−367 is 100 less than 921−267.

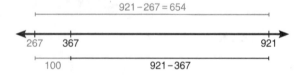

921−267 = 654. So, we have
921−367 = 654−100 = **554**.

141. We use 427−168 to find 427−148.

In 427−148, we start with the same number as in 427−168, but we subtract 20 less.

Subtracting 20 less makes our result greater by 20.

427−168 = 259. So, we have
427−148 = 259+20 = **279**.

— *or* —

We consider the difference on the number line.

148 is 20 units farther from 427 than 168 is from 427. So, 427−148 is 20 more than 427−168.

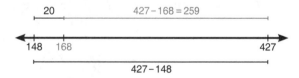

427−168 = 259. So, we have
427−148 = 259+20 = **279**.

142. We use 723−237 to find 823−137.

In 823−137, we start with 100 more and subtract 100 less than in 723−237.

Starting with 100 more makes our result greater by 100. Subtracting 100 less makes our result greater by 100 as well.

So, 823−137 is 100+100 = 200 more than 723−237.

723−237 = 486. So, we have
823−137 = 486+200 = **686**.

— *or* —

We consider the difference on the number line.

823 is 100 more than 723, and 137 is 100 less than 237. So, 823 and 137 are 100+100 = 200 units farther apart than 723 and 237.

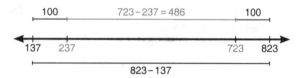

723−237 = 486. So, we have
823−137 = 486+200 = **686**.

143. We use 632−466 to find 622−476.

In 622−476, we start with 10 less and subtract 10 more than in 632−466.

Starting with 10 less makes our result smaller by 10. Subtracting 10 more makes our result smaller by 10 as well.

So, 622−476 is 10+10 = 20 less than 632−466.

632−466 = 166. So, we have
622−476 = 166−20 = **146**.

— *or* —

We consider the difference on the number line.

622 is 10 less than 632, and 476 is 10 more than 466. So, 622 and 476 are 10+10 = 20 units closer together than 632 and 466.

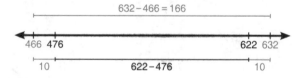

632−466 = 166. So, we have
622−476 = 166−20 = **146**.

144. 159 is 20 less than 179. If we start with the same amount and take away 20 less, the result will be 20 more.

So, 538−159 is **20** greater than 538−179.

— *or* —

538 and 159 are 20 units farther apart than 538 and 179 on the number line.

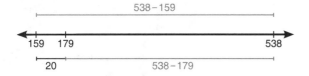

So, 538−159 is **20** greater than 538−179.

145. 319 is 400 less than 719. If we start with a number that is 400 less and take away the same amount, the result will be 400 less.

So, 319−144 is **400** less than 719−144.

— *or* —

319 and 144 are 400 units closer together than 719 and 144 on the number line.

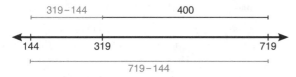

So, 319−144 is **400** less than 719−144.

146. The amount of time that passes does not change the difference between Greg's age and Meg's age. Over time, they both age the same amount! So, in 8 years, Greg will still be **13** years older than Meg.

147. Adding 6 to the big number and subtracting 6 from the small number makes the two resulting numbers 6+6 = 12 units farther apart.

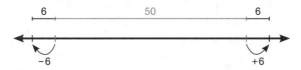

So, if the original numbers had a difference of 50, the new numbers have a difference of 50+12 = **62**.

148. When Polly gives 3 grapes to Molly, Polly loses 3 grapes *and* Molly gains 3 grapes. This decreases how many more grapes Polly has than Molly by 3+3 = 6.

Polly started with 10 more grapes than Molly. So, Polly now has 10−6 = **4** more grapes than Molly.

We can check with an example. If Polly had 25 grapes and Molly had 15 grapes, then after giving 3 grapes to Molly, Polly will have 22 grapes and Molly will have 18 grapes. This is a difference of 22−18 = 4. ✓

149. The smallest two-digit number is 10. Since 90−10 = 80, 10 and 90 is one pair of two-digit numbers with a difference of 80.

Increasing both 10 and 90 by 1 does not change the difference: 91−11 also equals 80. So, 11 and 91 is another pair of two-digit numbers with a difference of 80.

We can continue adding 1 to each number to get the following pairs that have a difference of 80:

10 and 90, 11 and 91, 12 and 92, 13 and 93, 14 and 94, 15 and 95, 16 and 96, 17 and 97, 18 and 98, 19 and 99.

The next pair, 20 and 100, does not contain two 2-digit numbers. So, there are **10** pairs of two-digit numbers with a difference of 80.

150. We can add the same amount to 85 and 38 without changing the difference. So, the difference between 85 and 38 is the same as the difference between 85+2 and 38+2.

So, 85−38 has the same difference as **87**−40.

151. 85−38 = 87−40 = **47**.

152. The difference between 271 and 94 is the same as the difference between 271+6 and 94+6.

So, 271−94 has the same difference as **277**−100.

153. 271−94 = 277−100 = **177**.

154. The difference between 183 and 57 is the same as the difference between 183+3 and 57+3.

So, 183−57 has the same difference as **186**−60.

155. 183−57 = 186−60 = **126**.

156. We consider each choice.

240 is 1 less than 241, and 81 is 1 less than 82. So, 240−81 has the same difference as 241−82. ✓

250 is 9 more than 241, and 91 is 9 more than 82. So, 250−91 has the same difference as 241−82. ✓

259 is 18 more than 241, and 100 is 18 more than 82. So, 259−100 has the same difference as 241−82. ✓

239 is 2 *less* than 241, but 84 is 2 *more* than 82. So, 239−84 is not the same difference as 241−82. ✗

200 is 41 less than 241, and 41 is 41 less than 82. So, 200−41 has the same difference as 241−82. ✓

   239−84

(240−81) (250−91) (259−100) 239−84 (200−41)

In each problem below, we subtract from left to right.

157. 32−12−2
= 20−2
= **18**

158. 148−116−16
= 32−16
= **16**

159. 297−143−43
= 154−43
= **111**

160. 201−99−40
= 102−40
= **62**

161.

$188-44-33-11$
$= 144-33-11$
$= 111-11$
$= \mathbf{100}$

162.

$285-25-25-25$
$= 260-25-25$
$= 235-25$
$= \mathbf{210}$

163.

$168-68-70-29$
$= 100-70-29$
$= 30-29$
$= \mathbf{1}$

164.

$345-45-67-89$
$= 300-67-89$
$= 233-89$
$= \mathbf{144}$

SUBTRACTION
All at Once
34

165. After her purchases, Alisha has $189-27-23$ dollars left.

Spending 27 dollars then spending 23 dollars leaves the same number of dollars as spending $27+23 = 50$ dollars all at once.

So, $189-27-23 = 189-50 = 139$. So, Alisha has **139** dollars left.

166. Snorg has $262-75-25$ minutes of his movie left to watch.

Watching 75 minutes then watching 25 minutes leaves the same number of minutes as watching $75+25 = 100$ minutes all at once.

So, $262-75-25 = 262-100 = 162$. So, Snorg has **162** minutes left to watch.

167. Taking away 38 then taking away 12 is the same as taking away $38+12 = 50$ all at once.

So, $84-38-12 = \boxed{84} - \boxed{50}$.

168. $84-38-12 = 84-50 = \mathbf{34}$.

169. Taking away 22, then taking away 33, then taking away 44 is the same as taking away $22+33+44 = 99$ all at once.

So, $511-22-33-44 = \boxed{511} - \boxed{99}$.

170. $511-22-33-44 = 511-99$
$= 511-100+1$
$= \mathbf{412}$.

SUBTRACTION
Part by Part
35

171. After Alfred takes 45 jelly beans, there are $137-45$ jelly beans left in the bag.

To make the subtraction easier, we subtract the 37 jelly beans Alfred gives to Sam, then subtract the 8 jelly beans Alfred eats.

$$137-45 = 137-37-8$$
$$= 100-8$$
$$= 92.$$

So, there are **92** jelly beans left in the bag.

172. After 167 birds fly away, there are $352-167$ birds still in the tree.

To make the subtraction easier, we subtract the 152 black birds that fly away, then subtract the 15 blue birds that fly away.

$$352-167 = 352-152-15$$
$$= 200-15$$
$$= 185.$$

So, there are **185** birds still in the tree.

173. We consider each choice.

Taking away 55 then taking away 33 is the same as taking away 88 all at once. So, $755-88 = 755-55-33$. ✓

Taking away 55 then adding 33 takes away less than 55 all together. So, $755-55+33$ is not equal to $755-88$. ✗

Taking away 5 then taking away 83 is the same as taking away 88 all at once. So, $755-88 = 755-5-83$. ✓

Taking away 44 then adding 44 is the same as doing nothing. So, $755-44+44$ is not equal to $755-88$. ✗

$\boxed{755-55-33}$ $755-55+33$ $\boxed{755-5-83}$ $755-44+44$

174. Since $86 = 74+12$, taking away 86 is the same as taking away 74, then taking away 12 more.

So, $174-86 = \boxed{174} - \boxed{74} - \boxed{12}$.

175. $174-86 = 174-74-12$
$= 100-12$
$= \mathbf{88}$.

176. Since $66 = 11+55$, taking away 66 is the same as taking away 11, then taking away 55 more.

So, $311-66 = \boxed{311} - \boxed{11} - \boxed{55}$.

177. $311-66 = 311-11-55$
$= 300-55$
$= \mathbf{245}$.

SUBTRACTION
Different Order
36

178. Picking 35 apples, then picking 82 apples leaves the same number of apples as picking 82 apples, then picking 35 apples. So, $182-35-82 = 182-82-35$.

$182-82 = 100$, and $100-35 = 65$. So, there are **65** apples still on the tree.

179. Juan has $117-45-17-35$ dollars left.

It does not matter what order Juan spent his money in. So, we can reorder the numbers we subtract. Since 17 is easier to subtract from 117 than 45 is, we write $117-45-17-35$ as $117-17-45-35$.

$117-17-45-35 = 100-45-35$. Taking away 45 then taking away 35 is the same as taking away 80 all at once. So, $100-45-35 = 100-80 = 20$.

So, Juan has **20** dollars left.

180. We change the order of the numbers being subtracted to make our computation easier:

$$71 - 45 - 11 = 71 - 11 - 45$$
$$= 60 - 45$$
$$= \mathbf{15}.$$

181. We change the order of the numbers being subtracted to make our computation easier:

$$465 - 9 - 365 = 465 - 365 - 9$$
$$= 100 - 9$$
$$= \mathbf{91}.$$

182. We change the order of the numbers being subtracted to make our computation easier:

$$234 - 150 - 34 = 234 - 34 - 150$$
$$= 200 - 150$$
$$= \mathbf{50}.$$

183. We change the order of the numbers being subtracted to make our computation easier:

$$225 - 37 - 75 = 225 - 75 - 37$$
$$= 150 - 37$$
$$= \mathbf{113}.$$

SUBTRACTION
Challenge Problems   37-39

184. We notice that each number is double the number to its right. Subtracting from left to right, we have

$$128 - 64 - 32 - 16 - 8 - 4 - 2 - 1$$
$$= \quad 64 - 32 - 16 - 8 - 4 - 2 - 1$$
$$= \quad\quad 32 - 16 - 8 - 4 - 2 - 1$$
$$= \quad\quad\quad 16 - 8 - 4 - 2 - 1$$
$$= \quad\quad\quad\quad 8 - 4 - 2 - 1$$
$$= \quad\quad\quad\quad\quad 4 - 2 - 1$$
$$= \quad\quad\quad\quad\quad\quad 2 - 1$$
$$= \quad\quad\quad\quad\quad\quad\quad \mathbf{1}.$$

185. After removing 654 pounds of sand, 321 pounds of sand remain. So, the amount of sand that was originally in the truck is the number that fills the blank below:

$$\_\_\_ - 654 = 321.$$

We can rewrite this subtraction problem using addition:

$$321 + 654 = \_\_\_.$$

So, there were $321 + 654 = \mathbf{975}$ pounds of sand in the truck before any sand was removed.

186. The four largest two-digit numbers are 99, 98, 97, and 96. So, Genie's result is $400 - 99 - 98 - 97 - 96$.

To subtract 99, we subtract 100, then add 1.
To subtract 98, we subtract 100, then add 2.
To subtract 97, we subtract 100, then add 3.
To subtract 96, we subtract 100, then add 4.

All together, we subtract four hundreds (400), and add a total of $1 + 2 + 3 + 4 = 10$. So, Genie's result is $400 - 99 - 98 - 97 - 96 = 400 - 400 + 10 = \mathbf{10}$.

187. Taking away the three missing numbers one by one is the same as taking away the sum of the three missing numbers all at once.

Since $40 - \underline{36} = 4$, we know that the sum of the three missing numbers is 36. So, we look for the number we add three copies of to get 36:

$$\_\_\_ + \_\_\_ + \_\_\_ = 36.$$

Since $\underline{12} + \underline{12} + \underline{12} = 36$, the missing number is 12.

$$40 - \underline{\ 12\ } - \underline{\ 12\ } - \underline{\ 12\ } = 4.$$

188. Since Emily is counting by 7's, we know:

- The 31st number is 7 more than the 30th.
- The 32nd number is $7 + 7 = 14$ more than the 30th.
- The 33rd number is $7 + 7 + 7 = 21$ more than the 30th.

So, the difference between the 33rd number and the 30th number is **21**.

In fact, the 30th number Emily says is 210, and the 33rd number she says is 231. We have $231 - 210 = 21$. But, we did not need to know these numbers!

189. We can rewrite $84 - 2\square = 5\square$ using addition:

$$5\square + 2\square = 84.$$

Adding 5 tens and 2 tens gives 7 tens, or 70.
Adding the ones digits gives $\square + \square$. So, we have

$$70 + \square + \square = 84.$$

Since $70 + \underline{7} + \underline{7} = 84$, the missing digit is 7.

$$84 - 2\boxed{7} = 5\boxed{7}$$

190. To make the result as large as possible, we want to start with the largest number possible and take away the two smallest numbers possible.

The largest three-digit number is 999.
The two smallest three-digit numbers are 100 and 101.

So, the largest possible result is $999 - 100 - 101 = \mathbf{798}$.

191. We can write any subtraction problem as an addition problem. So, if $\blacksquare - \triangle = \bigcirc$, then

$$\bigcirc + \triangle = \blacksquare. \checkmark$$

Since we can add numbers in any order, we also have

$$\triangle + \bigcirc = \blacksquare. \checkmark$$

We know that the sum of $\bigcirc$ and $\triangle$ is $\blacksquare$. So, we can draw the following picture:

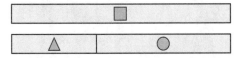

From this picture, we see that taking away $\bigcirc$ from $\blacksquare$ leaves $\triangle$. So, the following statement is true:

$$\blacksquare - \bigcirc = \triangle. \checkmark$$

We can use the picture to rule out each of the remaining statements:

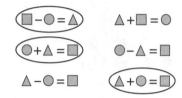

$\triangle + \square = \bigcirc$. ✗
$\bigcirc - \triangle = \square$. ✗
$\triangle - \bigcirc = \square$. ✗

We circle the three true statements.

$\boxed{\square - \bigcirc = \triangle}$ $\triangle + \square = \bigcirc$

$\boxed{\bigcirc + \triangle = \square}$ $\bigcirc - \triangle = \square$

$\triangle - \bigcirc = \square$ $\boxed{\triangle + \bigcirc = \square}$

As a check, we can replace each symbol with a number.

For example, suppose that $\square = 3$, $\triangle = 2$, and $\bigcirc = 1$. We first check that $\square - \triangle = \bigcirc$ is true: $3 - 2 = 1$. ✓

Then, we check each of the six choices.

$\square - \bigcirc = \triangle$ gives $3 - 1 = 2$. ✓
$\triangle + \square = \bigcirc$ gives $2 + 3 = 1$. ✗
$\bigcirc + \triangle = \square$ gives $1 + 2 = 3$. ✓
$\bigcirc - \triangle = \square$ gives $1 - 2 = 3$. ✗
$\triangle - \bigcirc = \square$ gives $2 - 1 = 3$. ✗
$\triangle + \bigcirc = \square$ gives $2 + 1 = 3$. ✓

192. The smaller the number that we subtract from 456, the larger the result will be. So, taking away the smallest number possible gives a result that is as large as possible.

Since the result is a two-digit number, the largest possible result is 99. So, we are looking for the number we subtract from 456 to get 99:

$$456 - \underline{\quad} = 99.$$

We can rewrite this as the easier subtraction problem below:

$$456 - 99 = \underline{\quad}.$$

Since $456 - 99 = \underline{357}$, we have $456 - \underline{357} = 99$.

So, **357** is the smallest number we can subtract from 456 to get a two-digit result.

193. The stool's height plus Stuart's height equals Stuart's dad's height.

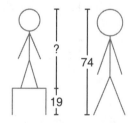

So, Stuart is $74 - 19 = 55$ inches tall.

This means that Stuart is $55 - 19 = \mathbf{36}$ inches taller than the stool.

194. From least to greatest, we list all six numbers that can be made using the digits 1, 2, and 3 once each:

$$123, \ 132, \ 213, \ 231, \ 312, \ 321.$$

The difference between the largest and the smallest number in Winnie's list is $321 - 123$. To make this computation easier, we subtract in parts:

$$321 - 123 = 321 - 121 - 2$$
$$= 200 - 2$$
$$= \mathbf{198}.$$

195. There are many ways to get the calculator to read 91. For example, we could push the $+10$ button nine times to get to 90, then the $+1$ button once. This is a total of 10 button presses.

Or, we could push the $+100$ button once, then the -1 button nine times to get to 91. This is also a total of 10 button presses.

We are looking for the *smallest* number of presses. So, with each press, we use the button that gets us closest to 91.

We start by pushing the $+100$ button to get to 100. This leaves us just 9 away from 91.

Next, we push the -10 button to get to 90. Now, we are just 1 away from 91.

Finally, we press the $+1$ to get to 91.

This is a total of **3** button presses.

There is no way to get 91 using 2 or fewer button presses.

EXPRESSIONS
Evaluating — 41

1.
$$7+3-5$$
$$= 10-5$$
$$= \mathbf{5}.$$

2.
$$19-8+11$$
$$= 11+11$$
$$= \mathbf{22}.$$

3.
$$34-13+9$$
$$= 21+9$$
$$= \mathbf{30}.$$

4.
$$275-204+111$$
$$= 71+111$$
$$= \mathbf{182}.$$

5.
$$70+50-12$$
$$= 120-12$$
$$= \mathbf{108}.$$

6.
$$200-79+55$$
$$= 121+55$$
$$= \mathbf{176}.$$

7.
$$19-16+12-14$$
$$= 3+12-14$$
$$= 15-14$$
$$= \mathbf{1}.$$

8.
$$59+30-17+13$$
$$= 89-17+13$$
$$= 72+13$$
$$= \mathbf{85}.$$

9.
$$580-30-140+60$$
$$= 550-140+60$$
$$= 410+60$$
$$= \mathbf{470}.$$

10.
$$203+177-70-70$$
$$= 380-70-70$$
$$= 310-70$$
$$= \mathbf{240}.$$

EXPRESSIONS
Cross-Number Puzzles — 42-43

11.

1	+	2	+	3	=	**6**
+		+		+		+
8	−	3	−	3	=	**2**
+		−		+		−
5	−	2	+	3	=	**6**
=		=		=		=
14	−	**3**	−	**9**	=	**2**

12.

20	+	12	−	5	=	**27**
−		−		+		−
6	+	3	+	5	=	**14**
+		−		−		+
8	−	4	+	2	=	**6**
=		=		=		=
22	+	**5**	−	**8**	=	**19**

13.

45	−	34	+	14	=	**25**
−		−		+		−
15	+	25	−	19	=	**21**
−		+		−		+
6	+	10	−	5	=	**11**
=		=		=		=
24	+	**19**	−	**28**	=	15

14.

15	+	120	+	70	=	**205**
+		−		−		−
45	+	98	+	17	=	**160**
+		+		−		+
39	−	11	+	23	=	**51**
=		=		=		=
99	−	**33**	+	**30**	=	**96**

15.

10	−	3	+	**2**	=	9
−		+		+		−
4	+	7	−	6	=	5
+		+		−		+
12	−	8	+	**2**	=	2
=		=		=		=
18	−	**18**	+	**6**	=	**6**

16.

11	+	**22**	+	6	=	39
−		−		+		−
8	+	**9**	−	**16**	=	1
+		−		−		+
15	−	1	−	**5**	=	9
=		=		=		=
18	+	12	+	**17**	=	**47**

EXPRESSIONS
Word Problems — 44-45

17. To find the sum of 3, 4, and 5, we add all three numbers: $3+4+5$.

$\boxed{3+4+5}$ $3-4+5$ $3+4-5$ $3-4-5$

18. Herman has 8 cookies, minus the 3 he ate, minus the 4 he gave to Edna. So, he has 8−3−4 cookies.

8+3+4   8−3+4   (8−3−4)   8+3−4

19. Starting with 45 balls in the basket, we subtract the 35 balls Lynn hit, then add the 33 that she put back, giving 45−35+33 golf balls.

45+35+33   45+35−33

(45−35+33)   45−35−33

20. To find 3 *less than* the sum of 4 and 5, we first find the sum of 4 and 5, then take away 3.

The sum of 4 and 5 is 4+5. Then, 3 less than this sum is 4+5−3.

4+5+3   3−4+5   (4+5−3)   5−4−3

21. Starting with 17 blocks, we add the 10 that Winnie placed on top, then subtract the 7 that fell off. This gives a tower that is **17+10−7 = 20** blocks tall.

22. Starting with 27 bow ties, we subtract the 15 Alex sells, then add the 6 he buys. This gives **27−15+6 = 18** bow ties.

23. We only care about the number of birds, not their colors.

We start with 36+43 birds. Then, we subtract the 19 that fly away. Finally, we add the 30 that land. This gives **36+43−19+30 = 90** birds in the tree.

24. To find 4 *more than* the difference of 9 and 6, we first find the difference of 9 and 6, then add 4.

The difference of 9 and 6 is 9−6. Then, 4 more than this difference is **9−6+4 = 7**.

25. To find 3 *less than* the sum of 5 and 8, we first find the sum of 5 and 8, then take away 3.

The sum of 5 and 8 is 5+8. Then, 3 less than this sum is **5+8−3 = 10**. You could have also written **8+5−3 = 10**.

Expression Search 46−47

We show one way to find the missing entries. You many have solved each in a different order.

26. We consider the 8 in the upper-left corner. The six expressions we can make using this 8 are listed below.

8
8+1
8+1+7
8+1+7+4
8+2
8+2+7

Of these, only 8+1 equals 9. So, we circle 8+1.

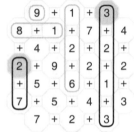

The only expression that uses the 9 in the upper-left corner is 9 itself.

We then consider the 1 in the top row. There are five expressions we can make using this 1, as shown below.

1
1+3
1+2
1+2+6
1+2+6+2

Of these, only 1+2+6 equals 9. So, we circle 1+2+6.

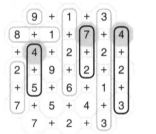

Then, the only expression using the shaded 3 that equals 9 is 3+2+1+3.

The only expression using the shaded 2 that equals 9 is 2+7.

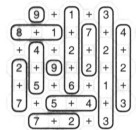

We continue circling expressions that equal 9 to complete the puzzle, as shown.

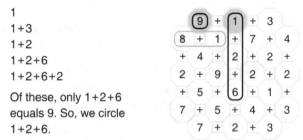

27. We consider the 6 in the upper-left corner. The six expressions we can make using this 6 are listed below.

6
6+36
6+36−16
6+7
6+7+7
6+7+7−36

Of these, only 6+7+7 equals 20. So, we circle 6+7+7.

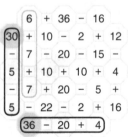

Then, the only expression using the shaded 30 that equals 20 is 30−5−5.

The only expression using the shaded 36 that equals 20 is 36−20+4.

We continue circling expressions that equal 20 to
complete the puzzle, as shown.

Step 3:

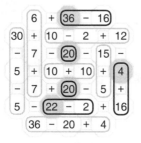

Step 4:

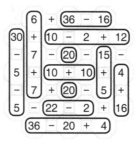

Step 5:

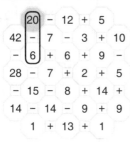

Final:

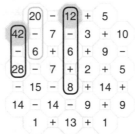

Step 3:

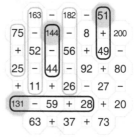

Final:

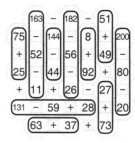

28. Step 1:

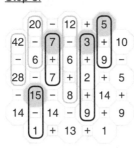

Step 2:

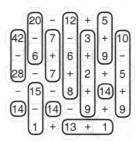

Step 3:

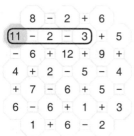

Final:

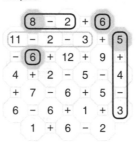

29. Step 1:

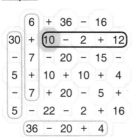

Step 2:

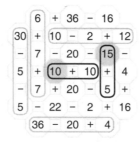

30. Step 1:

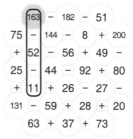

Step 2:

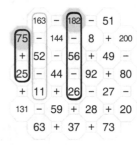

Step 3:

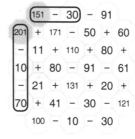

Final:

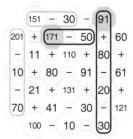

31. Step 1:

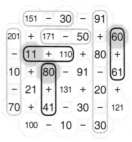

Step 2:

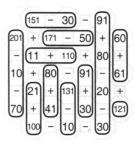

Step 3:

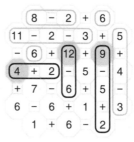

Final:

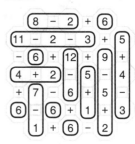

32. Step 1:   Step 2:

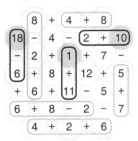   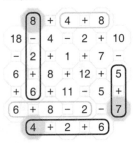

Step 3:   Final:

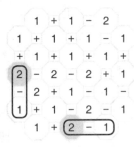   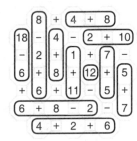

33. We first group the 2's with other numbers to create expressions that equal 1, then circle the remaining 1's.

Step 1:   Step 2:

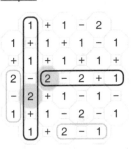

Step 3:   Final:

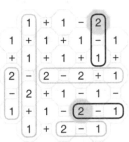   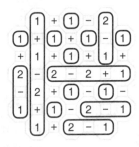

34. $23\ \boxed{-}\ 2 = 21$.

35. $15\ \boxed{+}\ 6\ \boxed{+}\ 7 = 28$.

36. Starting with 30, we must subtract 11 to get 19. So, we subtract both 9 and 2. This gives $30\ \boxed{-}\ 9\ \boxed{-}\ 2 = 19$.

37. $25\ \boxed{-}\ 25\ \boxed{+}\ 50 = 50$.

38. Starting with 90, we must add 2 more than we subtract to get 92. So, we add 36 and subtract 34.

This gives $90\ \boxed{+}\ 36\ \boxed{-}\ 34 = 92$.

39. Starting with 90, we must subtract 2 more than we add to get 88. So, we subtract 36 and add 34.

This gives $90\ \boxed{-}\ 36\ \boxed{+}\ 34 = 88$.

40. Starting with 20, we must add 1 more than we subtract to get 21. So, we add both 6 and 3, and subtract 8.

So, $20\ \boxed{+}\ 6\ \boxed{+}\ 3\ \boxed{-}\ 8 = 21$.

41. Starting with 300, we must subtract 100 to get 200. Subtracting 30, 30, and 40 is the same as subtracting 100.

So, $300\ \boxed{-}\ 30\ \boxed{-}\ 30\ \boxed{-}\ 40 = 200$.

42. We try starting with $18\ \boxed{-}\ 16$, which gives 2. There is no way to add or subtract 14 and 12 to get from 2 to 8. So, we try $18\ \boxed{+}\ 16$, which is 34. From 34, we can get 8 by subtracting 14 and 12.

So, $18\ \boxed{+}\ 16\ \boxed{-}\ 14\ \boxed{-}\ 12 = 8$.

43. We try starting with $38\ \boxed{+}\ 28$, which gives 66. There is no way to add or subtract 25 and 15 to get from 66 to 20. So, we try $38\ \boxed{-}\ 28$, which is 10. From 10, we can get 20 by adding 25 and subtracting 15.

So, $38\ \boxed{-}\ 28\ \boxed{+}\ 25\ \boxed{-}\ 15 = 20$.

44. To fill the first two boxes of $\square - \square + \square = 20$, we take a smaller number away from a larger number. This leaves three choices:

$18 - 17 + 19 = 20$ ✓
$19 - 17 + 18 = 20$ ✓
$19 - 18 + 17 = 18$ ✗

So, $\boxed{18} - \boxed{17} + \boxed{19} = 20$ and $\boxed{19} - \boxed{17} + \boxed{18} = 20$.

45. To fill the three boxes of $\square - \square - \square = 1$, we subtract two numbers from the first number. So, we take the two smaller numbers, 9 and 19, from the largest number, 29.

Then, there are two ways to order 9 and 19.

So, $\boxed{29} - \boxed{19} - \boxed{9} = 1$ and $\boxed{29} - \boxed{9} - \boxed{19} = 1$.

46. To fill the first two boxes of $\square - \square = \square + 11$, we take a smaller number away from a larger number. This leaves three choices.

$\boxed{44} - \boxed{22}$ and $\boxed{11} + 11$ both equal 22. ✓
$\boxed{44} - \boxed{11}$ and $\boxed{22} + 11$ both equal 33. ✓
$\boxed{22} - \boxed{11}$ and $\boxed{44} + 11$ are not equal. ✗

So, $\boxed{44} - \boxed{22} = \boxed{11} + 11$ and $\boxed{44} - \boxed{11} = \boxed{22} + 11$.

47. There are six ways to fill the boxes.

$\boxed{14} + 15 - \boxed{16} = \boxed{17}$. ✗ $\boxed{14} + 15 - \boxed{17} = \boxed{16}$. ✗
$\boxed{16} + 15 - \boxed{14} = \boxed{17}$. ✓ $\boxed{16} + 15 - \boxed{17} = \boxed{14}$. ✓
$\boxed{17} + 15 - \boxed{14} = \boxed{16}$. ✗ $\boxed{17} + 15 - \boxed{16} = \boxed{14}$. ✗

Only $\boxed{16} + 15 - \boxed{14} = \boxed{17}$ and $\boxed{16} + 15 - \boxed{17} = \boxed{14}$ make true statements.

48. We first evaluate the part inside the parentheses, then evaluate the resulting expression.

$$(8-2)+6$$
$$= 6+6$$
$$= 12.$$

So, $(8-2)+6 = \mathbf{12}$.

49. We first evaluate the part inside the parentheses, then evaluate the resulting expression.

$$8-(2+6)$$
$$= 8-8$$
$$= 0.$$

So, $8-(2+6) = \mathbf{0}$.

50. There are no parentheses, so we evaluate the expression from left to right.

$$8-2+6$$
$$= 6+6$$
$$= 12.$$

So, $8-2+6 = \mathbf{12}$.

51.
$$9-(5-3)$$
$$= 9-2$$
$$= \mathbf{7}.$$

52.
$$(9-5)-3$$
$$= 4-3$$
$$= \mathbf{1}.$$

53.
$$9-5-3$$
$$= 4-3$$
$$= \mathbf{1}.$$

54.
$$(7+4)-1$$
$$= 11-1$$
$$= \mathbf{10}.$$

55.
$$7+(4-1)$$
$$= 7+3$$
$$= \mathbf{10}.$$

56.
$$7+4-1$$
$$= 11-1$$
$$= \mathbf{10}.$$

57.
$$13-12+(7-6)$$
$$= 13-12+1$$
$$= 1+1$$
$$= \mathbf{2}.$$

58.
$$13-(12+7-6)$$
$$= 13-(19-6)$$
$$= 13-13$$
$$= \mathbf{0}.$$

59.
$$28-(11-5)-2$$
$$= 28-6-2$$
$$= 22-2$$
$$= \mathbf{20}.$$

60.
$$28-(11-5-2)$$
$$= 28-(6-2)$$
$$= 28-4$$
$$= \mathbf{24}.$$

61. $15-6+4 = 5$ is not true. Placing parentheses around the first two or three numbers does not change what we evaluate first.

There is only one other way to place parentheses:
$15-(6+4) = 15-10 = 5$.

So, $15-(6+4) = 5$.

62. $20-(4-3) = 19$.

63. We check ways we can place parentheses in $25-3+2+5$ that change what we evaluate first.

There are two ways to group two numbers:

$25-(3+2)+5$ gives $25-5+5 = 25$.
$25-3+(2+5)$ gives $25-3+7 = 29$.

There is one way to group three numbers:

$25-(3+2+5)$ gives $25-10 = 15$.

From our choices, only $25-(3+2)+5 = 25$ makes a true statement.

64. $70-50-(30-10) = 0$.

65. $45-(18-17+9) = 35$.

66. $10-(2+3)-1 = 4$.

67. $100-(45+25)+15 = 45$.

68. $17+6-(5+4) = 14$.

69. We check ways we can place parentheses in $7+6-5+4-3$ that change what we evaluate first.

There are three ways to group two numbers:

$7+(6-5)+4-3$ gives $7+1+4-3 = 9$.
$7+6-(5+4)-3$ gives $7+6-9-3 = 1$.
$7+6-5+(4-3)$ gives $7+6-5+1 = 9$.

There are two ways to group three numbers:

$7+(6-5+4)-3$ gives $7+5-3 = 9$.
$7+6-(5+4-3)$ gives $7+6-6 = 7$.

There is one way to group four numbers:

$7+(6-5+4-3)$ gives $7+2 = 9$.

From our choices, only $7+6-(5+4)-3 = 1$ makes a true statement.

70. We check ways we can place parentheses in $15-4+6-3+2$ that change what we evaluate first.

There are three ways to group two numbers:

$15-(4+6)-3+2$ gives $15-10-3+2 = 4$.
$15-4+(6-3)+2$ gives $15-4+3+2 = 16$.
$15-4+6-(3+2)$ gives $15-4+6-5 = 12$.

There are two ways to group three numbers:

$15-(4+6-3)+2$ gives $15-7+2 = 10$.
$15-4+(6-3+2)$ gives $15-4+5 = 16$.

There is one way to group four numbers:

$15-(4+6-3+2)$ gives $15-9 = 6$.

From our choices, only $15-(4+6-3)+2 = 10$ makes a true statement.

71.
$$10-(5-4)-(3+2)$$
$$= 10-1-5$$
$$= 9-5$$
$$= \mathbf{4}.$$

72.
$$1-(1-(1-1))$$
$$= 1-(1-0)$$
$$= 1-1$$
$$= \mathbf{0}.$$

73.
$$30-3-(30-(10+3))$$
$$= 30-3-(30-13)$$
$$= 30-3-17$$
$$= 27-17$$
$$= \mathbf{10}.$$

74.
$$24-(12-(3+3)-3)$$
$$= 24-(12-6-3)$$
$$= 24-(6-3)$$
$$= 24-3$$
$$= \mathbf{21}.$$

75.
$$39-(26-(7+9-5))$$
$$= 39-(26-(16-5))$$
$$= 39-(26-11)$$
$$= 39-15$$
$$= \mathbf{24}.$$

76.
$$4-(6-(8-(12-10)))$$
$$= 4-(6-(8-2))$$
$$= 4-(6-6)$$
$$= 4-0$$
$$= \mathbf{4}.$$

77. Alex starts with 12 red balloons and 7 green balloons. So, he has a total of $(12+7)$ balloons.

After 4 of these balloons pop, Alex has $(12+7)-4$ balloons left.

$$12-(7+4) \qquad (12-7)-4 \qquad 12-(7-4)$$
$$(12-7)+4 \qquad \boxed{(12+7)-4}$$

78. Lizzie has 23 Beastbucks. After buying the hula hoop, Grogg has $(38-16)$ Beastbucks. Together, Lizzie and Grogg have $23+(38-16)$ Beastbucks.

$$\boxed{23+(38-16)} \qquad 38-(23+16) \qquad 38-(23-16)$$
$$23+38+16 \qquad (38-23)+16$$

79. Ralph has 25 centipillars. After 9 caterpillars crawl away, Cammie has $(15-9)$ centipillars. So, Ralph has $25-(15-9)$ more centipillars than Cammie.

$$(25-9)-15 \qquad 25-15-9 \qquad (25-15)-9$$
$$\boxed{25-(15-9)} \qquad 25+(15-9)$$

80. Winnie starts with 64 ounces of juice in the pitcher. To fill four 12-ounce cups, she pours $12+12+12+12$ ounces of juice.

So, after she pours out a total of $(12+12+12+12)$ ounces of juice, there are $64-(12+12+12+12)$ ounces of juice left in the pitcher.

$$(64-4)-12 \qquad (64-12)+12+12+12$$
$$\boxed{64-(12+12+12+12)} \qquad 64-(4+12)$$

EXPRESSIONS

Review 54-55

81. We change one sign at a time and evaluate.

$10 \ominus 4-2+5-6-1 = 2.$ ✗
$10+4 \oplus 2+5-6-1 = 14.$ ✗
$10+4-2 \ominus 5-6-1 = 0.$ ✓
$10+4-2+5 \oplus 6-1 = 22.$ ✗
$10+4-2+5-6 \oplus 1 = 12.$ ✗

So, we circle the + in front of the 5 that must be changed to a − to make the expression equal zero.

$$10 + 4 - 2 \; \textcircled{+} \; 5 - 6 - 1$$

— or —

We evaluate the original expression.
$10+4-2+5-6-1 = 10.$ We want to change one sign so that this expression is 10 less than its current value.

If we subtract 5 instead of adding 5, the expression will be 10 less than its current value.

We check: $10+4-2-5-6-1 = 0.$ ✓

So, we circle the + in front of the 5 that must be changed to a − to make the expression equal zero.

$$10 + 4 - 2 \; \textcircled{+} \; 5 - 6 - 1$$

82. We evaluate each expression.

$6+12-1-5 = 12.$
$12+1-5-6 = 2.$
$12+5+1-6 = 12.$
$5-1+6+12 = 22.$
$12-6-5-1 = 0.$

$6+12-1-5$ and $12+5+1-6$ both equal 12.

$$\boxed{6+12-1-5} \qquad 12+1-5-6 \qquad \boxed{12+5+1-6}$$
$$5-1+6+12 \qquad 12-6-5-1$$

83. The difference of four and two is $4-2$. Nine minus this difference is $9-(4-2)$.

Evaluating, we have $9-(4-2) = 9-2 = 7.$

So, nine minus the difference of four and two is **7**.

84. The total team score of Griff and Cliff is $(15+21)$ points.

The total team score of Snark and Clark is $(19+11)$ points.

So, Griff and Cliff beat Snark and Clark by $(15+21)-(19+11)$ points.

$$\boxed{(15+21)-(19+11)} \qquad (21-15)+(19-11)$$
$$19-(15+(21-19)) \qquad 21+11-(19-15)$$

85. Without parentheses, $16-8-4+2$ equals 6.

We check ways we can place parentheses that change what we evaluate first.

There are three ways to place one pair of parentheses:

$16-(8-4)+2 = 14.$
$16-8-(4+2) = 2.$
$16-(8-4+2) = 10.$

There is one way to place two pairs of parentheses:

$16-(8-(4+2)) = 14.$

We cannot place a third pair of parentheses to change the order of what we evaluate first.

So, the smallest value we can get by adding parentheses to the expression $16-8-4+2$ is $16-8-(4+2) = \mathbf{2}.$

86. In the previous problem, we listed all of the different results we can get by adding parentheses to the expression $16-8-4+2$. The largest value we can get is $16-(8-4)+2 = \mathbf{14}$ or $16-(8-(4+2)) = \mathbf{14}.$

87. We work from the outside in.

Since $19-\underline{14} = 5$, we know the expression $(11+(8-\underline{}))$ is equal to 14. So, $(11+(8-\underline{})) = 14.$

Since $11+\underline{3} = 14$, we know the expression $(8-\underline{})$ is equal to 3. So, $(8-\underline{}) = 3.$

Since $8-\underline{5} = 3$, the number that fills the blank is **5**.

We check our work.
$19-(11+(8-5)) = 19-(11+3) = 19-14 = 5.$ ✓

88. We focus on the 15, since it is much larger than the other numbers. Subtracting 3 and 5 from 15 gives $15-3-5=7$.

We can also get 7 by adding the first three numbers: $4+2+1=7$.

So, $4\boxed{+}2\boxed{+}1\boxed{=}15\boxed{-}3\boxed{-}5$.

89. We evaluate each expression by replacing the ▲ with 3.

▲$+2=3+2=$**5**.
▲$-1=3-1=$**2**.
$4+$▲$=4+3=$**7**.
$6-$▲$=6-3=$**3**.

90. We evaluate each expression by replacing each ◆ with 9.

◆$+$◆$=9+9=$**18**.
◆$-7+$◆$=9-7+9=$**11**.
$20-$◆$-2=20-9-2=$**9**.
$20-($◆$-5)-($◆$+5)=20-(9-5)-(9+5)=20-4-14=$**2**.

91. We evaluate each expression by replacing each ■ with 99.

■$-98+$■$=99-98+99=$**100**.
$400-$■$-$■$-$■$=400-99-99-99=$**103**.
$($■$+2)-($■$-1)=(99+2)-(99-1)=101-98=$**3**.
■$-3+$■$-2+$■$-1=99-3+99-2+99-1=$**291**.

92. We evaluate the expression by replacing the ☾ with 11.

☾$+10=11+10=$**21**.

93. We evaluate the expression by replacing each ◯ with 6.

◯$+$◯$+$◯$=6+6+6=$**18**.

94. We evaluate the expression by replacing the ☾ with 11 and the ◯ with 6.

☾$+$◯$=11+6=$**17**.

95. ☾$-($◯$+1)=11-(6+1)=11-7=$**4**.

96. $18-($◯$+$☾$)=18-(6+11)=18-17=$**1**.

97. ◯$+3-($☾$-10)=6+3-(11-10)=6+3-1=$**8**.

98. $($☾$+$◯$)-($☾$-$◯$)=(11+6)-(11-6)=17-5=$**12**.

99. ◯$+$☾$-($◯$+5)=6+11-(6+5)=6+11-11=$**6**.

100. When the temperature is 45 degrees, ✹$=45$. So, ✹-15 is $45-15=30$, and a cup of lemonade costs **30** cents.

When the temperature is 68 degrees, ✹$=68$. So, ✹-15 is $68-15=53$, and a cup of lemonade costs **53** cents.

When the temperature is 85 degrees, ✹$=85$. So, ✹-15 is $85-15=70$, and a cup of lemonade costs **70** cents.

When the temperature is 99 degrees, ✹$=99$. So, ✹-15 is $99-15=84$, and a cup of lemonade costs **84** cents.

101. For a monster that has 4 feet, ✿$=4$.
So, ✿$+$✿$+$✿$+2$ is $4+4+4+2=14$, and a monster with 4 feet should pack **14** socks for ski camp.

For a monster that has 5 feet, ✿$=5$.
So, ✿$+$✿$+$✿$+2$ is $5+5+5+2=17$, and a monster with 5 feet should pack **17** socks for ski camp.

For a monster that has 10 feet, ✿$=10$.

So, ✿$+$✿$+$✿$+2$ is $10+10+10+2=32$, and a monster with 10 feet should pack **32** socks for ski camp.

For a monster that has 50 feet, ✿$=50$.
So, ✿$+$✿$+$✿$+2$ is $50+50+50+2=152$, and a monster with 50 feet should pack **152** socks for ski camp.

102. When ▲$=5$ and ■$=8$, we have
▲$-(10-$■$)=5-(10-8)=5-2=$**3**.

When ▲$=25$ and ■$=3$, we have
▲$-(10-$■$)=25-(10-3)=25-7=$**18**.

When ▲$=2$ and ■$=9$, we have
▲$-(10-$■$)=2-(10-9)=2-1=$**1**.

When ▲$=17$ and ■$=7$, we have
▲$-(10-$■$)=17-(10-7)=17-3=$**14**.

103. For a monster that read 7 extra chapters in 2 weeks,
●$=7$ and ◆$=2$. So, ●$+$●$-$◆ is $7+7-2=12$, and Trish earns **12** Baxter Bucks.

For a monster that read 3 extra chapters in 4 weeks,
●$=3$ and ◆$=4$. So, ●$+$●$-$◆ is $3+3-4=2$, and Brian earns **2** Baxter Bucks.

For a monster that read 5 extra chapters in 10 weeks,
●$=5$ and ◆$=10$. So, ●$+$●$-$◆ is $5+5-10=0$, and Dobbin earns **0** Baxter Bucks. Sorry, Dobbin.

104. For a team that has 5 goals, 4 saves, and 1 foul,
★$=5$, ●$=4$, and ■$=1$. So, ★$+$★$+$●$-$■ is $5+5+4-1=13$, and the team scores **13** points.

For a team that has 11 goals, 2 saves, and 9 fouls,
★$=11$, ●$=2$, and ■$=9$. So, ★$+$★$+$●$-$■ is $11+11+2-9=15$, and the team scores **15** points.

For a team that has 0 goals, 6 saves, and 4 fouls,
★$=0$, ●$=6$, and ■$=4$. So, ★$+$★$+$●$-$■ is $0+0+6-4=2$, and the team scores **2** points.

105. Doubling ◯ donuts gives ◯$+$◯ donuts. So, if Globb makes ◯ donuts, Robb makes ◯$+$◯ donuts.

For example, if Globb makes 6 donuts, Robb makes $6+6$ donuts.

◯   ◯$+2$ ◯$+$◯$+$◯   2   (◯$+$◯)

106. To get from Hana's age to Frida's age, we add 3 to Hana's age. So, if Hana's age is ▮▮, Frida's age is ▮▮$+3$.

For example, if Hana is 10, then Frida is $10+3=13$.

▮▮   ▮▮-3 (▮▮$+3$)   ▮▮$+$▮▮$+$▮▮ $3-$▮▮

107. Since Captain Kraken spends ✿ coins, he has $100-$✿ coins left.

For example, if Kraken spends 15 coins, he has $100-15$ coins left.

✿   ($100-$✿) $100+$✿   ✿$-(100-$✿$)$ ✿-100

108. Since the team has ★ players, and ☾ are drivers, then there are ★$-$☾ team members who are not drivers.

For example, if the team has 5 players, and 2 are drivers, then there are $5-2$ team members who are not drivers.

☾$+$★   ☾$-$★   ☾$-($★$-$☾$)$ (★$-$☾)   (★$-$☾$)+$☾

109. The number of girls in Ms. Franz's class is double the number of boys. So, if there are ● boys, there are ●+● girls.

The total number of students in Ms. Franz's class is the sum of the number of boys and the number of girls. So, the total number of students in the class is ●+(●+●).

For example, if there are 10 boys, there are 10+10 girls, and 10+(10+10) total students in the class.

    ●+●   ●−(●−●)   (●+(●+●))

    ●+(●−●)   ●−(●+●)

110. Tayvon is 5 inches taller than Stephanie. So, Stephanie is 5 inches shorter than Tayvon.

If Tayvon is **T** inches tall, Stephanie is **T**−5 inches tall.

For example, if Tayvon is 45 inches tall, Stephanie is 45−5 inches tall.

(**T**−5)   5−**T**   **T**+5   **T**+(**T**−5)   **T**−(**T**+5)

111. The first koalaphant has ▲ tusks.

The second koalaphant has ▲ tusks.

The third koalaphant has ▲ tusks.

The fourth koalaphant has ▲ tusks.

So, the total number of tusks in a group of 4 koalaphants is ▲+▲+▲+▲.

For example, if koalaphants have 7 tusks, then 4 koalaphants have 7+7+7+7 tusks.

    ▲−4   ▲+4   ▲+(▲−4)

    ▲+(▲+4)   (▲+▲+▲+▲)

112. To find the number of push-ups, we subtract the total number of sit-ups and jumping jacks from 50.

So, if a team member does **S** sit-ups and **J** jumping jacks, she must do 50−(**S**+**J**) push-ups.

For example, if a team member does 10 sit-ups and 15 jumping jacks, she must do 50−(10+15) push-ups.

    50−**S**+**J**   50−(**S**−**J**)   50+(**S**+**J**)

    50+(**S**−**J**)   (50−(**S**+**J**))

EXPRESSIONS

Simplifying   62–64

113. Subtracting any number from itself gives zero. So, 28−28 = **0**.

114. Subtracting any number from itself gives zero. So, ◆−◆ = **0**.

115. 333−10+10 = 323+10 = **333**.

         — *or* —

Subtracting a number and then adding the same number is the same as doing nothing. So, 333−10+10 = **333**.

116. Subtracting a number and then adding the same number is the same as doing nothing. So, ■−45+45 = **■**.

117. Subtracting any number from itself gives zero. So, 65−65+54 = 0+54 = **54**.

118. Subtracting any number from itself gives zero. So, ♣−♣+11 = 0+11 = **11**.

119. Subtracting any number from itself gives zero. So, 75−(57−57) = 75−0 = **75**.

120. Subtracting any number from itself gives zero. So, ▲−(●−●) = ▲−0 = **▲**.

121. 19−(19−1) = 19−18 = **1**.

         — *or* —

(19−1) is 1 less than 19. So, the difference between 19 and (19−1) is 1. We have 19−(19−1) = **1**.

122. ☾ is 3 more than (☾−3). So, the difference between ☾ and (☾−3) is 3. We have ☾−(☾−3) = **3**.

123. We can add numbers in any order. So, to simplify ●+5+5+5+5, we first add the four 5's. Since 5+5+5+5 = 20, we have ●+5+5+5+5 = ●+**20**.

*(You may have written this as **20+●**.)*

124. Subtracting any number from itself gives zero. So, (♣−♣) = 0 and (■−■) = 0. We have ▲+(♣−♣)+(■−■) = ▲+0+0 = **▲**.

125. Subtracting any number from itself gives zero. So, ◗−◗+◗+7 = 0+◗+7 = ◗+**7**.

*(You may have written this as **7+◗**.)*

126. We can add numbers in any order. So, 18+☾+32 = 18+32+☾ = 50+**☾**.

*(You may have written this as **☾+50**.)*

127. In 20−○−10, we take away ○ and 10 from 20. We can take away ○ and 10 in either order. So, 20−○−10 = 20−10−○ = **10−○**.

128. (▼+5) is 5 more than ▼. Also, (▼+2) is 2 more than ▼. So, the difference between (▼+5) and (▼+2) is 5−2 = 3.

We have (▼+5)−(▼+2) = **3**.

         — *or* —

We consider the difference on the number line.

▼+2 is 2 units to the right of ▼, and ▼+5 is 5 units to the right of ▼.

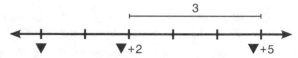

So, ▼+2 and ▼+5 are 3 units apart. This means that (▼+5)−(▼+2) = **3**.

129. We try replacing the ◆ in the expression ■−(■−◆) with a few numbers.

When ◆ = 1, the expression is ■−(■−1). (■−1) is 1 less than ■. So, the difference between ■ and (■−1) is 1.

When ◆ = 20, the expression is ■−(■−20). (■−20) is 20 less than ■. So, the difference between ■ and (■−20) is 20.

So, we know that
($\blacksquare$−$\blacklozenge$) is $\blacklozenge$ less than $\blacksquare$. So, the difference between $\blacksquare$ and ($\blacksquare$−$\blacklozenge$) is $\blacklozenge$.

So, $\blacksquare$−($\blacksquare$−$\blacklozenge$)=$\blacklozenge$.

— *or* —

We consider the difference on the number line.

$\blacksquare$−$\blacklozenge$ is $\blacklozenge$ units to the left of $\blacksquare$.

So, the difference between $\blacksquare$ and $\blacksquare$−$\blacklozenge$ is $\blacklozenge$. This means that $\blacksquare$−($\blacksquare$−$\blacklozenge$)=$\blacklozenge$.

130. We first simplify $\blacktriangledown$−19+19−19.

Subtracting 19 then adding 19 is the same as doing nothing. So, $\blacktriangledown$−19+19−19=$\blacktriangledown$+0−19=$\blacktriangledown$−19.

We evaluate $\blacktriangledown$−19 for each value of $\blacktriangledown$.

When $\blacktriangledown$=20, we have $\blacktriangledown$−19=20−19=**1**.
When $\blacktriangledown$=25, we have $\blacktriangledown$−19=25−19=**6**.
When $\blacktriangledown$=49, we have $\blacktriangledown$−19=49−19=**30**.
When $\blacktriangledown$=87, we have $\blacktriangledown$−19=87−19=**68**.

131. We first simplify 100−$\bigstar$−$\bigstar$+$\bigstar$. Subtracting a number then adding it is the same as doing nothing.

So, 100−$\bigstar$−$\bigstar$+$\bigstar$=100−$\bigstar$+0=100−$\bigstar$.

When $\bigstar$=15, we have 100−$\bigstar$=100−15=**85**.
When $\bigstar$=5, we have 100−$\bigstar$=100−5=**95**.
When $\bigstar$=33, we have 100−$\bigstar$=100−33=**67**.
When $\bigstar$=49, we have 100−$\bigstar$=100−49=**51**.

132. We first simplify $\maltese$+($\bigcirc$−$\bigcirc$)+($\bigcirc$−$\bigcirc$)+$\bigcirc$. Subtracting any number from itself gives zero. So,

$$\maltese+(\bigcirc-\bigcirc)+(\bigcirc-\bigcirc)+\bigcirc=\maltese+0+0+\bigcirc$$
$$=\maltese+\bigcirc.$$

When $\maltese$=6 and $\bigcirc$=8, we have $\maltese$+$\bigcirc$=6+8=**14**.
When $\maltese$=20 and $\bigcirc$=10, we have $\maltese$+$\bigcirc$=20+10=**30**.
When $\maltese$=33 and $\bigcirc$=27, we have $\maltese$+$\bigcirc$=33+27=**60**.
When $\maltese$=111 and $\bigcirc$=86, we have $\maltese$+$\bigcirc$=111+86=**197**.

133. We first simplify 15+($\blacksquare$−$\leftmoon$)−($\blacksquare$−$\leftmoon$)+$\blacksquare$.
Adding ($\blacksquare$−$\leftmoon$) then subtracting ($\blacksquare$−$\leftmoon$) is the same as doing nothing. So, 15+($\blacksquare$−$\leftmoon$)−($\blacksquare$−$\leftmoon$)+$\blacksquare$=15+$\blacksquare$.

Since $\leftmoon$ does not appear in the simplified expression, its value does not matter here.

When $\blacksquare$=61, we have 15+$\blacksquare$=15+61=**76**.
When $\blacksquare$=143, we have 15+$\blacksquare$=15+143=**158**.
When $\blacksquare$=212, we have 15+$\blacksquare$=15+212=**227**.
When $\blacksquare$=800, we have 15+$\blacksquare$=15+800=**815**.

EXPRESSIONS *Equations* 65

134. We have 1 ⊕ 2 ⊜ 3.

135. We try placing an equals sign in each circle.

When 1 ⊜ 3 ◯ 5 ◯ 1, there is no way to fill the circles in 3 ◯ 5 ◯ 1 with + and − signs to get a result of 1.

When 1 ◯ 3 ⊜ 5 ◯ 1, on the left side, we can make

1+3 = 4. On the right side, we can make 5−1 = 4. ✓

When 1 ◯ 3 ◯ 5 ⊜ 1, there is no way to fill the circles in 1 ◯ 3 ◯ 5 with + and − signs to get a result of 1.

So, we place the equals sign in the center circle, and place the other signs so that both sides equal 4.

$$1 \oplus 3 \circleddash 5 \ominus 1.$$

We use this strategy to find the solutions for the following problems.

136. 4 ⊕ 5 ⊕ 6 ⊜ 15.

137. 1 ⊜ 3 ⊕ 2 ⊖ 4.

You may also have filled in the circles as shown below.
1 ⊖ 3 ⊜ 2 ⊖ 4.

Both sides of this equation are negative! You'll learn about negative numbers in Beast Academy 4C.

138. 20 ⊜ 40 ⊖ 15 ⊖ 5.

139. 25 ⊖ 24 ⊕ 26 ⊜ 27.

You may also have filled in the circles as shown below.
25 ⊜ 24 ⊖ 26 ⊕ 27.

You'll learn how to compute 24−26+27 in Beast Academy 2C.

140. 16 ⊜ 18 ⊖ 14 ⊕ 12.

You may also have filled in the circles as shown below.
16 ⊖ 18 ⊕ 14 ⊜ 12.

You'll learn how to compute 16−18+14 in Beast Academy 2C.

141. 3 ⊕ 1 ⊜ 8 ⊖ 1 ⊖ 3.

EXPRESSIONS *Mismo* 66-68

We show one way to find the missing entries of these puzzles. You may have solved each in a different order.

142. There is only one way to make an equation in each row:

2 ⊕ 4 ⊜ 6 and 5 ⊜ 6 ⊖ 1.

So, we fill the rows as shown.

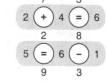

We check the columns.

7 ⊕ 2 ⊜ 9. ✓

5 ⊜ 8 ⊖ 3. ✓

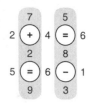

143. There are two ways to make an equation in the bottom row:

15 ⊜ 7 ⊕ 8 and 15 ⊖ 7 ⊜ 8.

But, there is only one way to make an equation in the top row:

9 ⊜ 15 ⊖ 6.

So, we fill the top row as shown.

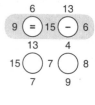

We complete the puzzle as shown.

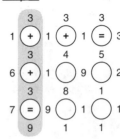

We use the strategy discussed in the previous solutions to solve the Mismo puzzles that follow.

144. Step 1:

Final:

145. Step 1:

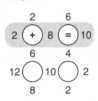

Final:

146. Each row and column has two ways to make an equation.

There are two ways to make an equation in the left column:

$15 \ (=) \ 9 \ (+) \ 6$ or $15 \ (-) \ 9 \ (=) \ 6$.

We try the first equation.

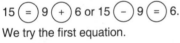

We cannot form an equation in the bottom row, since 11+3 is not equal to 8. So, we cannot use $15 \ (=) \ 9 \ (+) \ 6$ to solve the puzzle.

We try $15 \ (-) \ 9 \ (=) \ 6$, and finish the puzzle as shown.

Step 1:

Final:

We use the strategies discussed in the previous solutions to solve the Mismo puzzles that follow.

147. We complete the puzzle as shown.

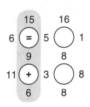

148. Step 1:

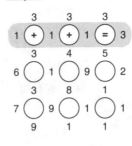

Step 2:

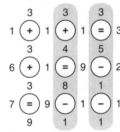

Step 3:

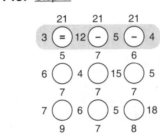

Final:

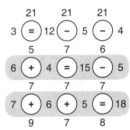

149. Step 1:

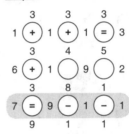

Final:

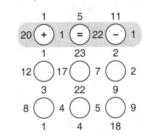

150. Step 1:

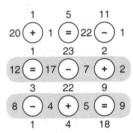

Final:

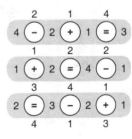

151. Step 1:

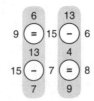

Final:

152. Step 1:

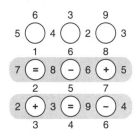

Final:

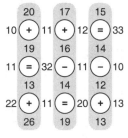

153. Step 1:

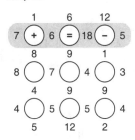

Final:

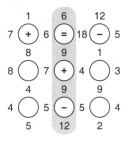

154. Step 1:

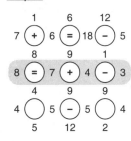

Step 2:

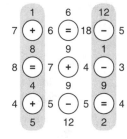

Step 3:

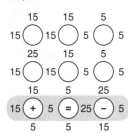

Final:

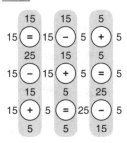

155. Step 1:

Final:

156. Since 6+5 = 11, we have ♦ = **6**.

157. Since 12+8 = 20, we have ✖ = **8**.

158. Since 50−40 = 10, we have ★ = **50**.

159. Since 16 = 116−100, we have ◯ = **116**.

160. Since 22 = 11+11, we have ● = **11**.

161. Since 50+1 = 51, we know that ✱+✱ equals 50.

$$✱+✱+1 = 51$$
$$50\ \ +1 = 51$$

Since 25+25 = 50, we have ✱ = **25**.

We replace ✱ with 25 to check our answer:
25+25+1 = 51. ✓

162. We guess values for ♦, and check to see if they work.

If ♦ = 10, then 24−♦ gives 14. This is not equal to ♦.

We need the right side of the equation to be larger, so we try larger values of ♦.

If ♦ = 11, then 24−♦ gives 13. This is not equal to ♦.

If ♦ = 12, then 24−♦ gives 12. This is equal to ♦.

So, ♦ = **12**.

Guessing can be a useful tool when solving problems! You'll learn more about guessing and checking in the next chapter.

— *or* —

Since taking away ♦ from 24 leaves ♦, adding ♦ to ♦ will give us 24.

Since 12+12 = 24, we have ♦ = **12**.

We replace ♦ with 12 to check our answer:
24−12 = 12. ✓

163. On both sides of the equation, we are adding ▲ to a number.

On the left side, we add ▲ to 5. On the right side, we add ▲ to ▲. So, ▲ and 5 are equal. This means ▲ = **5**.

We replace ▲ with 5 to check our answer:
5+▲ = 5+5 = 10, and ▲+▲ = 5+5 = 10. ✓

164. Since 13−8 = 5, we have (☾+3) = 8.

Then, since (5+3) = 8, we have ☾ = **5**.

We replace ☾ with 5 to check our answer:
13−(5+3) = 13−8 = 5. ✓

165. We guess values for ♣, and check to see if they work.

If ♣ = 10, then 24−♣ gives 14 and ♣+♣ gives 20. These amounts are not equal.

We need the right side of the equation to be smaller, so we try a smaller value of ♣.

If ♣ = 9, then 24−♣ gives 15 and ♣+♣ gives 18. These amounts are not equal.

If ♣ = 8, then 24−♣ gives 16 and ♣+♣ gives 16.

This works! The equation 24−♣ = ♣+♣ is true for ♣ = **8**.

166. We evaluate each of the six expressions.

$5+5+5 = 15.$ $5+4-3 = 6.$ $3+4-5 = 2.$
$2+2+2 = 6.$ $3-2+1 = 2.$ $4+5+6 = 15.$

We write an equation using each pair of equal expressions, as shown below.

$$3+4-5 = 3-2+1$$
$$5+4-3 = 2+2+2$$
$$5+5+5 = 4+5+6$$

You may have written these equations in a different order. In each equation, you may have also switched the left and right sides.

167. There are three different circles. We try placing an equals sign in each circle.

Placing an equals sign in the first circle, we have $101 \;\text{\textcircled{=}}\; 1 \;\text{\textcircled{+}}\; 99 \;\text{\textcircled{+}}\; 1.$

Placing an equals sign in the second circle, we have $101 \;\text{\textcircled{--}}\; 1 \;\text{\textcircled{=}}\; 99 \;\text{\textcircled{+}}\; 1.$

Placing an equals sign in the third circle, we have $101 \;\text{\textcircled{--}}\; 1 \;\text{\textcircled{--}}\; 99 \;\text{\textcircled{=}}\; 1.$

So, the three different equations are:

$101 \;\text{\textcircled{=}}\; 1 \;\text{\textcircled{+}}\; 99 \;\text{\textcircled{+}}\; 1,$
$101 \;\text{\textcircled{--}}\; 1 \;\text{\textcircled{=}}\; 99 \;\text{\textcircled{+}}\; 1,$ and
$101 \;\text{\textcircled{--}}\; 1 \;\text{\textcircled{--}}\; 99 \;\text{\textcircled{=}}\; 1.$

168. We check ways we can place parentheses that change what we evaluate first.

There are four ways to group two numbers.

$32-(16-8)-4-2-1$ gives 17. ✗
$32-16-(8-4)-2-1$ gives 9. ✗
$32-16-8-(4-2)-1$ gives 5. ✗
$32-16-8-4-(2-1)$ gives 3. ✗

There are three ways to group three numbers.

$32-(16-8-4)-2-1$ gives 25. ✗
$32-16-(8-4-2)-1$ gives 13. ✓
$32-16-8-(4-2-1)$ gives 7. ✗

There are two ways to group four numbers.

$32-(16-8-4-2)-1$ gives 29. ✗
$32-16-(8-4-2-1)$ gives 15. ✗

There is one way to group five numbers.

$32-(16-8-4-2-1)$ gives 31. ✗

From our choices, only $32-16-(8-4-2)-1 = 13$ makes a true statement.

169. Starting with 5, we look for ways to group the subtraction to create an expression that equals 3.

Since $5-1-1 = 3$, we first look for a way to group the subtraction that gives us $5-1-1$.
There is only one way to do this:
$5-(4-3)-(2-1) = 5-1-1 = 3.$

Since $5-2 = 3$, we next look for a way to group the subtraction that gives us $5-2$.
There is only one way to do this:
$5-(4-(3-2)-1) = 5-(4-1-1) = 5-2 = 3.$

So, the two different ways to place two pairs of parentheses in the statement to make it true are:

$5-(4-3)-(2-1) = 3$ and $5-(4-(3-2)-1) = 3.$

170. The expression $(\leftmoon+10)-(\leftmoon-10)$ is the difference between two numbers, $(\leftmoon+10)$ and $(\leftmoon-10)$. We look at these two numbers on a number line.

$\leftmoon-10$ is ten units to the left of $\leftmoon$, and $\leftmoon+10$ is ten units to the right of $\leftmoon$.

The difference between $\leftmoon+10$ and $\leftmoon-10$ is the distance between them on the number line. That distance is $10+10 = 20$ units.

So, $(\leftmoon+10)-(\leftmoon-10)$ simplifies to **20**.

— *or* —

$(\leftmoon+10)$ is 10 more than $\leftmoon$, which is 10 more than $(\leftmoon-10)$.

So, $(\leftmoon+10)$ is 20 more than $(\leftmoon-10)$.

So, $(\leftmoon+10)-(\leftmoon-10) = $ **20**.

171. First, we look at $\blacktriangle+\bigstar = \blacktriangle+17$. To the left of the equals sign, we add $\bigstar$ to $\blacktriangle$. To the right of the equals sign, we add 17 to $\blacktriangle$.

So, $\bigstar$ and 17 are equal.

We then look at $\blacktriangle+\blacktriangle+\bigstar = \blacktriangle+39$. Replacing the $\bigstar$ with 17 gives $\blacktriangle+\blacktriangle+17 = \blacktriangle+39$.

We can also write this as $\blacktriangle+(\blacktriangle+17) = \blacktriangle+39$. To the left of the equals sign, we add $(\blacktriangle+17)$ to $\blacktriangle$. To the right of the equals sign, we add 39 to $\blacktriangle$.

So, $\blacktriangle+17 = 39$.

Since $\underline{22}+17 = 39$, we have $\blacktriangle = $ **22**.

We replace $\blacktriangle$ with 22 and $\bigstar$ with 17 in both equations to check our answer:

$\blacktriangle+\bigstar = 22+17 = 39$ and $\blacktriangle+17 = 22+17 = 39.$ ✓

$\blacktriangle+\blacktriangle+\bigstar = 22+22+17 = 61$ and $\blacktriangle+39 = 22+39 = 61.$ ✓

Guessing & Checking 73-75

1. If Sam is 10, then Tram would be $10+5 = $ **15**. The sum of their ages would be $10+15 = $ **25**.

2. If Sam is 15, then Tram would be $15+5 = $ **20**. The sum of their ages would be $15+20 = $ **35**.

3. We are given that the sum of Sam's and Tram's ages is 31. In the previous problem, we found that if Sam is 15, then the sum of Sam's and Tram's ages is 35, which is too big. So, Sam must be **younger** than 15.

4. We guess that Sam is 14. Then, Tram would be $14+5 = 19$. The sum of their ages would be $14+19 = 33$. Still a little too big. So, we guess that Sam is 13.

If Sam is 13, then Tram would be $13+5 = 18$. The sum of their ages would be $13+18 = 31$. This works!

So, Sam is **13**.

5. If Ned has 15 toes, then Ming would have $15+15 = $ **30** toes. Together, they would have $15+30 = $ **45** toes.

6. If Ned has 20 toes, then Ming would have $20+20 = $ **40** toes. Together, they would have $20+40 = $ **60** toes.

7. Together, Ned and Ming have 54 toes. Our guesses from the two previous problems tell us that Ned has between 15 and 20 toes.

If Ned has 18 toes, then Ming would have $18+18 = 36$ toes. Together, they would have $18+36 = 54$ toes. This works! So, Ned has **18** toes.

8. If Mike has 4 five-dollar bills, then he would need **5** ten-dollar bills to have a total of 9 bills. The total value of Mike's bills would be $5+5+5+5+10+10+10+10+10 = $ **70** dollars.

9. If Mike has 6 five-dollar bills, then he would need **3** ten-dollar bills to have a total of 9 bills. The total value would be $5+5+5+5+5+5+10+10+10 = $ **60** dollars.

10. Mike's bills are worth a total of $55. Our guesses from the previous problems tell us that Mike has more than 6 five-dollar bills.

If Mike has 7 five-dollar bills, then he would need 2 ten-dollar bills to have a total of 9 bills. The total value of Mike's bills would be $5+5+5+5+5+5+5+10+10 = 55$ dollars. This works!

So, Mike has **7** five-dollar bills.

11. __10__ + __10__ + __10__ $= 30$.

12. We start by guessing numbers that are easy to add, then adjust our guesses until we find the right answer.

__10__ + __10__ + __10__ + __10__ $= 40$. ✗ Too small.

__15__ + __15__ + __15__ + __15__ $= 60$. ✗ Too large.

__13__ + __13__ + __13__ + __13__ $= 52$. ✓ Got it!

13. We start by guessing numbers in the left blank that are easy to subtract, then adjust to find the right answer.

$78 - $ __30__ $= $ __30__ ? ✗ $78-30 = 48$. So, 30 is too small.

$78 - $ __40__ $= $ __40__ ? ✗ $78-40 = 38$. So, 40 is too big.

$78 - $ __39__ $= $ __39__ ? ✓ Got it!

14. We start by guessing numbers that are easy to work with.

__20__ $+16 = 54-$ __20__ ? ✗
$20+16 = 36$, but $54-20 = 34$. Close! We try 21.

__21__ $+16 = 54-$ __21__ ? ✗
$21+16 = 37$, but $54-21 = 33$.
Our first guess was better, so we try 19.

__19__ $+16 = 54-$ __19__ ? ✓ Got it!
$19+16 = 35$, and $54-19 = 35$.

15. We start by guessing numbers that are easy to work with.

__10__ $+$ __10__ $= 27-$ __10__ ? ✗
$10+10 = 20$, but $27-10 = 17$.

__9__ $+$ __9__ $= 27-$ __9__ ? ✓ Got it!
$9+9 = 18$, and $27-9 = 18$.

16. We start by guessing numbers that are easy to work with.

$30+$ __10__ $= 50-(4+$ __10__ $)$? ✗
$30+10 = 40$, but $50-(4+10) = 36$.

$30+$ __15__ $= 50-(4+$ __15__ $)$? ✗
$30+15 = 45$, but $50-(4+15) = 31$.

Our first guess was better, so we try a number smaller than 10.

$30+$ __8__ $= 50-(4+$ __8__ $)$? ✓ Got it!
$30+8 = 38$, and $50-(4+8) = 38$.

Guess & Check Challenge 76-77

17. We guess that the larger number is 10. For their difference to be 5, the smaller number must be $10-5 = 5$. Then, their sum is $10+5 = 15$. ✗ Too big.

We guess that the larger number is 9. For their difference to be 5, the smaller number must be $9-5 = 4$. Then, their sum is $9+4 = 13$. ✓ This works!

So, the larger of the two numbers is **9**.

18. We guess that Brenda has 3 nickels and 3 dimes. 3 nickels and 3 dimes are worth $5+5+5+10+10+10 = 45$ cents. ✗ Too high.

To decrease the value of Brenda's coins, we guess more nickels and fewer dimes.

We guess that Brenda has 4 nickels and 2 dimes. 4 nickels and 2 dimes are worth $5+5+5+5+10+10 = 40$ cents. ✓ This works!

So, Brenda has **4** nickels.

19. We guess that Mr. Hayfield has 5 chickens and 5 cows. These animals have $2+2+2+2+2+4+4+4+4+4 = 30$ legs. ✗ Too many legs.

To decrease the total number of legs, we try more chickens and fewer cows.

6 chickens and 4 cows have 28 legs. ✗ Still too high.

7 chickens and 3 cows have 26 legs. ✓ This works!

So, Mr. Hayfield has **3** cows.

20. Winnie's number is the sum of two digits. So, Winnie's number is no more than $9+9 = 18$.

Grogg's number is double Winnie's number. So, Grogg's number is no more than $18+18 = 36$.

We guess values of Grogg's number that are smaller than 36 and look for when Grogg's number is double Winnie's.

If Grogg's number is 35, 34, 33, 32, 31, or 30, Winnie's number will be less than 8, 7, 6, 5, 4, or 3. ✗ Too small.

Similarly, when we try numbers for Grogg in the 20's, Winnie's number is always too small.

If Grogg's number is 19, Winnie's number is $1+9 = 10$. ✗

If Grogg's number is 18, Winnie's number is $1+8 = 9$. ✓ Since $9+9 = 18$, this works! So, Grogg's number is **18**.

There is no other 2-digit number that works.

21. We guess values for △, and check to see if they work.

We guess $△ = 5$. From our first equation, we have $5+5+□ = 13$. Since $5+5+\underline{3} = 13$, we have $□ = 3$. We check these values in the second equation.

If $△ = 5$ and $□ = 3$, then $△+□+□$ gives $5+3+3 = 11$. But, the second equation tells us that $△+□+□$ is 17. So, △ cannot equal 5.

We try other values for △.

If $△ = 6$, then our first equation gives $□ = 1$. But, these values do not work in our second equation.

We continue to guess and check for values of △ until we find $△ = 3$ and $□ = 7$.

We check our work:
$3+3+7 = 13$. ✓
$3+7+7 = 17$. ✓
So, $△+□ = 3+7 = $ **10**.

— *or* —

We can solve the problem without guessing.

We know $△+△+□$ is 13 and $△+□+□$ is 17. So, adding $(△+△+□)$ and $(△+□+□)$ gives $13+17 = 30$.

So, $(△+△+□)+(△+□+□) = 30$.

We can add numbers in any order, so the parentheses do not matter.

Since 3 △'s and 3 □'s sum to 30, we know that 1 △ and 1 □ sum to 10.

So, $△+□ = $ **10**.

Note that we never needed to find the values of △ and □, since we were only asked for their sum.

22. We guess that the small bag holds 10 apples. Then, the big bag holds $10+10 = 20$ apples. Together, the bags hold $10+20 = 30$ apples. ✗ This is too few.

If the small bag holds 15 apples, then the big bag holds $15+15 = 30$ apples. Together, they hold $15+30 = 45$ apples. ✗ This is too many, but only by a little.

If the small bag holds 14 apples, then the big bag holds $14+14 = 28$ apples. Together, they hold $14+28 = 42$ apples. ✓ This works!

So, the small bag holds 14 apples, and the big bag holds 28 apples.

Next, we guess the number of apples we must move so that the bags hold the same number of apples.

If we move 5 apples, the big bag will hold $28-5 = 23$ apples, and the small bag will hold $14+5 = 19$ apples. ✗ We need to move more apples.

If we move 7 apples, the big bag will hold $28-7 = 21$ apples, and the small bag will hold $14+7 = 21$ apples. ✓ This works!

So, **7** apples must be moved so that both bags hold the same number of apples.

Sym-Sums   78-81

23. In the top row, we have $△+△+△ = 9$. Since $\underline{3}+\underline{3}+\underline{3} = 9$, we know $△ = 3$.

We use the other rows to find ○ and □.

We guess $○ = 1$. Since the sum of the middle row is 7, we have $□ = 3$.

If $○ = 1$ and $□ = 3$, then in the bottom row we have $□+○+○ = 3+1+1 = 5$.

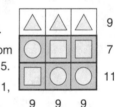

But, the bottom row has a sum of 11, not 5. So, ○ cannot be 1.

If $○ = 2$, then there is no whole number we can use for □ to make $○+□+□ = 7$ in the middle row. So, ○ cannot be 2.

We guess $○ = 3$. Since the sum of the middle row is 7, we have $□ = 2$. But, these values do not work in the bottom row. So, ○ cannot be 3.

If $○ = 4$, then there is no whole number we can use for □ to make $○+□+□ = 7$ in the middle row. So, ○ cannot be 4.

We guess $○ = 5$. Since the sum of the middle row is 7, we have $□ = 1$. If $○ = 5$ and $□ = 1$, then in the bottom row we have $□+○+○ = 1+5+5 = 11$. This works!

We check the other rows and columns to make sure each sum is correct. ✓

So, $△ = $ **3**, $○ = $ **5**, and $□ = $ **1**.

24. In the middle column, we have $\bigcirc + \bigcirc + \bigcirc = 6$. Since $\underline{2} + \underline{2} + \underline{2} = 6$, we know $\bigcirc = 2$.

We use the other columns to find $\triangle$ and $\square$.

We guess $\triangle = 4$. Since the sum of the left column is 10, we have $\square = 2$.

If $\triangle = 4$ and $\square = 2$, in the right column we have $\square + \square + \triangle = 2 + 2 + 4 = 8$.

But, the right column has a sum of 11, not 8. So, $\triangle$ cannot be 4.

$\triangle$	$\bigcirc$	$\square$	9
$\triangle$	$\bigcirc$	$\square$	9
$\square$	$\bigcirc$	$\triangle$	9
10	6	11	

We guess $\triangle = 3$. Since the sum of the left column is 10, we have $\square = 4$. If $\triangle = 3$ and $\square = 4$, in the right column we have $\square + \square + \triangle = 4 + 4 + 3 = 11$. This works!

We check the other rows and columns to make sure each sum is correct. ✓

So, $\triangle = \mathbf{3}$, $\bigcirc = \mathbf{2}$, and $\square = \mathbf{4}$.

25. We start with a row or column that only has two different shapes. In the left column, we have $\triangle + \triangle + \bigcirc = 4$. Since $2 + 2 = 4$, we know $\triangle$ cannot be more than 2. So, $\triangle$ is 0, 1, or 2.

We guess $\triangle = 0$. Since the sum of the left column is 4, we have $\bigcirc = 4$.

If $\triangle = 0$ and $\bigcirc = 4$, in the bottom row we have $\bigcirc + \triangle + \bigcirc = 4 + 0 + 4 = 8$.

But, the bottom row has a sum of 5, not 8. So, $\triangle$ cannot be 0.

$\triangle$	$\bigcirc$	$\square$	6
$\triangle$	$\square$	$\square$	7
$\bigcirc$	$\triangle$	$\bigcirc$	5
4	6	8	

We guess $\triangle = 1$. Since the sum of the left column is 4, we have $\bigcirc = 2$.

If $\triangle = 1$ and $\bigcirc = 2$, in the bottom row we have $\bigcirc + \triangle + \bigcirc = 2 + 1 + 2 = 5$. This works!

So, $\triangle = 1$ and $\bigcirc = 2$.

To find $\square$, we use the top row: $\triangle + \bigcirc + \square = 6$. This means $1 + 2 + \square = 6$. Since $1 + 2 + \underline{3} = 6$, we have $\square = 3$.

We check the other rows and columns to make sure each sum is correct. ✓

So, $\triangle = \mathbf{1}$, $\bigcirc = \mathbf{2}$, and $\square = \mathbf{3}$.

26. We guess and check using the strategies in the previous problem to find $\triangle = \mathbf{2}$, $\bigcirc = \mathbf{1}$, and $\square = \mathbf{4}$.

27. We start with a row or column that only has two different shapes. In the top row, we have $\bigcirc + \bigcirc + \triangle = 9$.

We guess $\bigcirc = 4$. Since the sum of the top row is 9, we have $\triangle = 1$.

In the middle column, we have $\bigcirc + \triangle + \triangle = 4 + 1 + 1 = 6$.

But, the middle column has a sum of 12, not 6. So, $\bigcirc$ cannot be 4.

$\bigcirc$	$\bigcirc$	$\triangle$	9
$\square$	$\triangle$	$\square$	11
$\square$	$\triangle$	$\bigcirc$	10
8	12	10	

We guess $\bigcirc = 2$. Since the sum of the top row is 9, we have $\triangle = 5$. In the middle column we have $\bigcirc + \triangle + \triangle = 2 + 5 + 5 = 12$. This works!

So, $\bigcirc = 2$ and $\triangle = 5$.

To find $\square$, we use the bottom row: $\square + \triangle + \bigcirc = 10$, which means $\square + 5 + 2 = 10$. Since $\underline{3} + 5 + 2 = 10$, we have $\square = 3$.

We check the other rows and columns to make sure each sum is correct. ✓

So, $\triangle = \mathbf{5}$, $\bigcirc = \mathbf{2}$, and $\square = \mathbf{3}$.

28. We guess and check using the strategies in the previous problems to find $\triangle = \mathbf{2}$, $\bigcirc = \mathbf{5}$, and $\square = \mathbf{4}$.

29. We guess and check using the strategies in the previous problems to find $\triangle = \mathbf{4}$, $\bigcirc = \mathbf{6}$, and $\square = \mathbf{2}$.

30. We guess and check using the strategies in the previous problems to find $\triangle = \mathbf{4}$, $\bigcirc = \mathbf{1}$, and $\square = \mathbf{5}$.

31. The numbers in the middle row have a small sum: $\bigcirc + \bigcirc + \triangle = 5$. This limits the possibilities for $\bigcirc$ and $\triangle$.

Either:
$\bigcirc = 0$ and $\triangle = 5$,
$\bigcirc = 1$ and $\triangle = 3$, or
$\bigcirc = 2$ and $\triangle = 1$.

$\square$	$\bigcirc$	$\triangle$	103
$\bigcirc$	$\bigcirc$	$\triangle$	5
$\triangle$	$\square$	$\square$	201
103	104	102	

We try these values in the other rows and columns to find $\square$. Only $\bigcirc = 2$, $\triangle = 1$, and $\square = 100$ works for every row and column.

So, $\triangle = \mathbf{1}$, $\bigcirc = \mathbf{2}$, and $\square = \mathbf{100}$.

— *or* —

In the middle row we have $\bigcirc + \bigcirc + \triangle = 5$. So, the values of $\bigcirc$ and $\triangle$ must be small.

The sum of the top row with 1 $\square$ is a little over 100. The sum of the bottom row with 2 $\square$'s is a little over 200.

Since $\bigcirc$ and $\triangle$ are small, the value of $\square$ must be about 100. We guess $\square = 100$.

If $\square = 100$, the bottom row gives us $\triangle = 1$ and the middle column gives us $\bigcirc = 2$.

We check the other rows and columns to make sure each sum is correct. ✓

So, $\triangle = \mathbf{1}$, $\bigcirc = \mathbf{2}$, and $\square = \mathbf{100}$.

32. We guess and check to find values that work. There are some clever observations that can help us guess well.

For example, we can compare the top two rows. In the top row, two $\triangle$'s plus one $\bigcirc$ equals 42. In the middle row, two $\triangle$'s plus one $\square$ equals 43. So, we know that $\square$ is 1 more than $\bigcirc$.

We can use this to make guesses for $\bigcirc$ and $\square$ in the right column, where $\bigcirc + \square + \square = 8$.

Since $\square$ is 1 more than $\bigcirc$, we guess $\bigcirc = 1$ and $\square = 2$.

This gives $\bigcirc + \square + \square = 1 + 2 + 2 = 5$.

But, the right column has a sum of 8. So, $\bigcirc$ is larger than 1.

$\triangle$	$\triangle$	$\bigcirc$	42
$\triangle$	$\triangle$	$\square$	43
$\square$	$\bigcirc$	$\square$	8
43	42	8	

We guess ◯ = 2 and ▢ = 3.

This gives ◯ + ▢ + ▢ = 2 + 3 + 3 = 8. This works!

We use these values to find △ = 20.

We check the other rows and columns to make sure each sum is correct. ✓

So, △ = **20**, ◯ = **2**, and ▢ = **3**

— or —

The rows and columns with two △'s have a sum that is a little over 40.

The rows and columns with no △'s have a sum of 8.

So, △ + △ is about 40. Since 20 + 20 = 40, we guess △ = 20.

If △ = 20, the top row gives ◯ = 2, and the second row gives ▢ = 3.

We check the other rows and columns to make sure each sum is correct. ✓

So, △ = **20**, ◯ = **2**, and ▢ = **3**.

33. We guess and check to find values that work. There are some clever observations that can help us guess well.

For example, we can compare the top two rows.

In the top row, two △'s plus one ▢ equals 50.
In the middle row, two △'s plus one ◯ equals 55.
So, we know that ◯ is 5 more than ▢.

We use this to help us guess and check.

We find △ = **20**, ◯ = **15**, and ▢ = **10**.

34. We guess and check to find values that work. There are some clever observations that help us guess well.

For example, we can compare the bottom row to the left column.

In the bottom row, two ◯'s plus one ▢ equals 20.
In the left column, two ◯'s plus one △ equals 22.
So, we know that △ is 2 more than ▢.

We use this to help us guess and check.

We find △ = **10**, ◯ = **6**, and ▢ = **8**.

35. Before Ted spent $20 on a video game, he had 45 + 20 = 65 dollars. Before he earned $12 doing chores, he had 65 − 12 = **53** dollars.

36. Before she ate 3, Mira had 18 + 3 = 21 cookies.
Before she bought 12, Mira had 21 − 12 = 9 cookies.
Before she sold 25, Mira had 9 + 25 = **34** cookies.

37. We start with Betsy, since we know her height is 45 inches.

Sheila is 2 inches taller than Betsy: 45 + 2 = 47 inches.
Marge is 8 inches shorter than Sheila: 47 − 8 = 39 inches.
Hildy is 5 inches taller than Marge: 39 + 5 = 44 inches.

So, Hildy is **44** inches tall.

38. Hector got off the elevator on floor 36.

To get to floor 36, the elevator went up 15 floors from floor 36 − 15 = 21.

To get to floor 21, the elevator went up 9 floors from floor 21 − 9 = 12.

To get to floor 12, the elevator went down 18 floors from floor 18 + 12 = 30.

To get to floor 30, the elevator went up 6 floors from floor 30 − 6 = 24, where Hector got on.

So, Hector's room is on floor **24**.

We check our work.

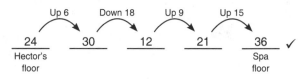

39. Level 4 is worth 600 points.

Since 300 + 300 = 600, level 3 is worth 300 points.

Since 150 + 150 = 300, level 2 is worth 150 points.

Since 75 + 75 = 150, level 1 is worth **75** points.

We check our work.

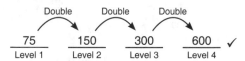

40. Every day, Skippy adds 12, then subtracts 5, then subtracts 1 from his acorn collection. So, he ends each day with 12 − 5 − 1 = 6 acorns more than he had the day before.

If Skippy ended Friday with 100 acorns,
he ended Thursday with 100 − 6 = 94 acorns,
he ended Wednesday with 94 − 6 = 88 acorns,
he ended Tuesday with 88 − 6 = 82 acorns, and
he ended Monday with 82 − 6 = **76** acorns.

We check our work.

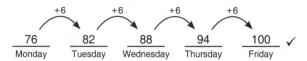

41. 25 + 2 = 27. Then, 27 − 5 = 22.

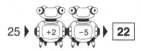

42. 14 − 6 = 8. Then, doubling 8 gives 8 + 8 = 16.

43. We work from right to left.

To get 35, we add 15 to 20.

To get 20, we subtract 9 from 29.

Check: 29−9+15 = 35. ✓

44. We work from right to left.

To get 20, we double 10.

To get 10, we add 5 to 5.

Check: 5+5 = 10, and doubling 10 gives 20. ✓

45. We work from right to left.

To get 9, we subtract 1 from 10.

To get 10, we double 5.

To get 5, we add 1 to 4.

Check: 4+1 = 5, doubling 5 gives 10, and 10−1 = 9. ✓

46. We work from right to left.

To get 9, we add 1 to 8.

To get 8, we double 4.

To get 4, we subtract 1 from 5.

Check: 5−1 = 4, doubling 4 gives 8, and 8+1 = 9. ✓

47. We work from right to left.

To get 99, we add 99 to 0.

To get 0, we subtract 99 from 99.

To get 99, we add 99 to 0.

To get 0, we subtract 99 from 99.

Check: 99−99+99−99+99 = 99. ✓

— *or* —

Subtracting 99 then adding 99 is the same as doing nothing. So, the first pair of bots cancel each other.

Similarly, the second pair of bots cancel each other.

So, we start with the same number we end with, **99**.

48. We work from right to left.

To get 2, we double 1.

To get 1, we subtract 3 from 4.

To get 4, we double 2.

To get 2, we subtract 4 from 6.

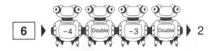

Check: 6−4 is 2. We double 2 to get 4. We subtract 3 to get 1. Finally, we double 1 to get 2. ✓

49. We work from right to left.

To get 50, we double 25.

To get 25, we subtract 9 from 34.

To get 34, we double 17.

To get 17, we subtract 9 from 26.

To get 26, we double 13.

Check: We double 13 to get 26. We subtract 9 to get 26−9 = 17. We double 17 to get 34. We subtract 9 to get 34−9 = 25. Finally, we double 25 to get 50. ✓

Knight Paths   86-87

50. There are *two* circles that the knight can move to on its first move.

But, working backwards, the knight can only reach the 5 from *one* of the unlabeled circles. We label this circle 4.

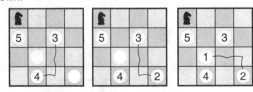

Continuing to work backwards, there is only one way the knight can reach each circle. We number the path as shown.

We check that our final path works. ✓

51. We work backwards from the circle marked 5 to find the only possible path.

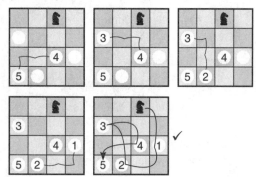

52. We work backwards from the circle marked 7 to find the only possible path.

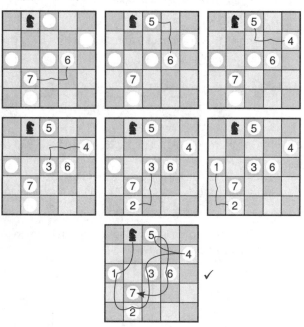

53. We work backwards from the circle marked 7 to find the only possible path.

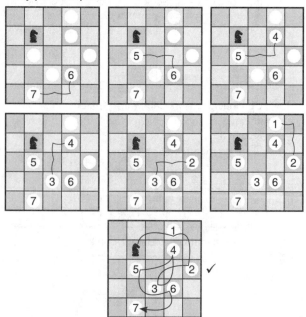

54. We work backwards from the circle marked 7.

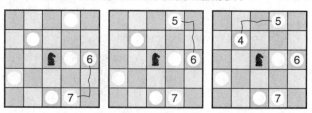

There are two circles we can reach the 4 from.

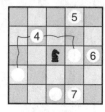

But, there is now only one unlabeled circle the knight can move to first. So, we finish the puzzle working forwards.

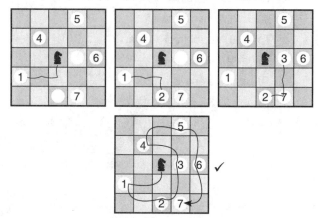

55. We work backwards from the circle marked 8.

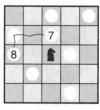

Then, there are two circles we can reach the 7 from.

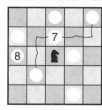

The circle in the top-right corner can only be reached from one blank circle. So, if we do not connect the top-right circle to the 7, it is a dead end. This means the top-right circle must be 6. We continue working backwards to label the 5.

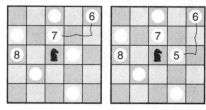

From here, there are two circles we can reach the 5 from, and 2 circles that the knight can reach on its first move. We guess and check to label the remaining circles on the path as shown below.

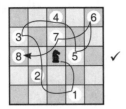

56. We are not given the last (8th) circle. But, we can find out which circle must be last on the path!

The circle below labeled 8 can only be reached from one other circle and cannot be the first circle the knight moves to. So, it must be the end of the knight's path.

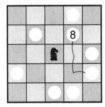

We continue working backwards to complete the path as shown.

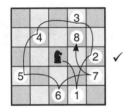

57. We are not given the last (12th) circle. But, we can find out which circle must be last on the path!

The circle below labeled 12 can only be reached from one other circle and cannot be the first circle the knight moves to. So, it must be the end of the knight's path. We label the last two circles as shown.

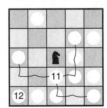

Then, there are three circles we can reach the 11 from.

The circle in the bottom-right corner can only be reached from one blank circle. So, if we do not connect the bottom-right circle to the 11, it is a dead end. This means

the bottom-right circle must be 10. We continue working backwards to label the remaining circles using the strategies from the previous puzzles.

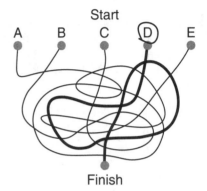

58. Starting at the *finish*, we trace the line to connect it to **D**.

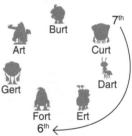

59. We work backwards around the circle, moving in the *opposite* direction that Zog went. Since Curt was the last (7th) moster to get a sticker, and Zog gave a sticker to every third monster, Fort was the 6th monster to get a sticker.

Continuing to work backwards, we see that Burt was 5th to get a sticker, Ert was 4th, Art was 3rd, Dart was 2nd, and **Gert** was 1st.

60. Since 60 is the sum of 27 and 33, we know that 60 was the last number Lizzie wrote.

So, before she wrote 60, Lizzie's paper looked like this:

21   26
6
33   15   27

Before she crossed out one of the four numbers above, Lizzie added two numbers to get 33. One of these numbers was 27. Since 27 + 6 = 33, the other number was 6. So, 6 was the last number above that Lizzie crossed out.

Before she crossed out 6, Lizzie's paper looked like this:

Since 33 is the sum of 6 and 27, we know that 33 was the last number Lizzie wrote.

Before she wrote 33, Lizzie's paper looked like this:

Before she crossed out one of the three numbers above, she added two uncrossed numbers to get 27. One of these numbers was 6. Since 6+21=27, the other number was 21. So, 21 was the last number above that Lizzie crossed out.

Before she crossed out 21, Lizzie's paper looked like this:

Since 27 is the sum of 21 and 6, we know that 27 was the last number Lizzie wrote.

Before she wrote 27, Lizzie's paper looked like this:

Before she crossed out one of the two numbers above, she added two uncrossed numbers to get 21. One of these numbers was 6. Since 6+15=21, the other number was 15.

Before she crossed out 15, Lizzie's paper looked like this:

Since 21 is the sum of 15 and 6, we know that 21 was the last number Lizzie wrote.

Before she wrote 21, Lizzie's paper looked like this:

So, the first number that Lizzie crossed out was 26, and the three numbers Lizzie started with were **6, 15,** and **26**.

— *or* —

The only three numbers that are not the sum of two other numbers are 6, 15, and 26. Since Lizzie could not have gotten any of those numbers by adding two other numbers, she must have started with them.

The three numbers that Lizzie started with were **6, 15,** and **26**.

61. **a.** We add the last number plus the digits of the number before it to get the next number in the list.

The number after 33 is 33+2+4 = 39, then 39+3+3 = 45, then 45+3+9 = 57, and finally 57+4+5 = 66.

15, 18, 24, 33, **39**, **45**, **57**, **66**

b. Since we are always adding from left to right, the numbers always get bigger from left to right.

We look for the number to the left of 29. Its digits get added to 29 to give 38. Since 29+9 = 38, the digits of the number left of 29 have a sum of 9.

The number left of 29 must be between 12 on the far left and 29 to its right. There are only 2 numbers between 12 and 29 that have a digit sum of 9. They are 18 and 27. We try 18.

12, ____, ____, ____, **18**, 29, 38, 49

Now, the number to the left of 18 has digits that we add to 18 to get 29. So, the digits of the number left of 18 have a sum of 11.

The smallest number that has a digit sum of 11 is 29, which is larger than 18. So, we cannot use 18 to the left of 29. So, we must use 27.

12, ____, ____, ____, **27**, 29, 38, 49

Now, the number to the left of 27 has digits that we add to 27 to get 29. So, the digits of the number left of 27 have a sum of 2. The only number between 12 and 27 that has a digit sum of 2 is 20.

12, ____, ____, **20**, **27**, 29, 38, 49

Left of 20, we need a number with digits that sum to 7. The only number between 12 and 20 that has a digit sum of 7 is 16.

12, ____, **16**, **20**, **27**, 29, 38, 49

Left of 16, we need a number with digits that sum to 4. The only number between 12 and 16 that has a digit sum of 4 is 13. We complete the DigiSum list as shown below and check our answer from left to right.

12, **13**, **16**, **20**, **27**, 29, 38, 49

62. We draw a circle to stand for each light.

We count **7** lights.

63. We draw 5 posts on each side of the field, including the post at each corner.

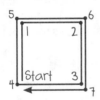

We count **16** posts.

64. We start by drawing a B for each of the 7 boys:

B   B   B   B   B   B   B

We place a girl (G) between each pair of boys, and a girl on each end so that every boy is between two girls:

G B G B G B G B G B G B G B G

Taking away any of the G's above leaves at least one boy who is not between two girls. So, this is the smallest number of girls that could be in the row.

We count **8** girls.

65. We draw Bronck's path around the square block, counting right turns as we go.

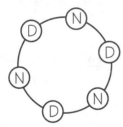

Bronck makes a total of **7** right turns.

66. Alex has a total of 6 nickels and dimes. We can arrange them in a circle as shown below.

There are 6 places we can place a penny so that no two pennies are next to each other.

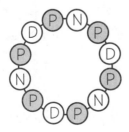

There is nowhere else we can place a penny on the circle that is not next to another penny. So, **6** is the largest number of pennies Alex can place.

We can arrange the nickels and dimes however we want, as long as the pennies separate them.

67. We can draw a few kids and look for a pattern.

With 2 kids, there is 1 hand-hold.

With 3 kids, there are 2 hand-holds.

With 4 kids, there are 3 hand-holds.

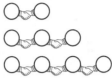

Each time we add another kid, we add 1 more hand-hold. So, there is always 1 more kid than there are hand-holds.

So, if there are 12 hand-holds, then there are **13** kids.

68. We draw a tree and use the measurements given to label the known distances as shown below.

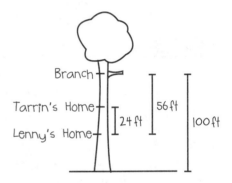

Lenny's home is 56 feet below the branch. So, Lenny's home is $100 - 56 = 44$ feet above the ground. We label this distance.

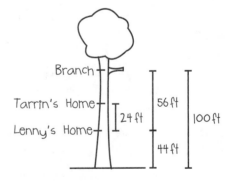

Tarrin's home is 24 feet above Lenny's home.
So, Tarrin's home is $44 + 24 = $ **68** feet above the ground.

69. We start with a number line, with marks drawn on it for every foot. We label Amy (A) and Ben (B) 5 feet from each other on the line.

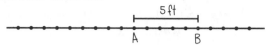

Next, Kenneth (K) and Amy are 9 feet apart. We can draw Kenneth to the right or the left of Amy.

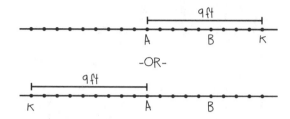

Finally, we place Deepa (D) 6 feet from Kenneth in our drawing. We try to place Deepa as close to Ben as possible in both drawings as shown below.

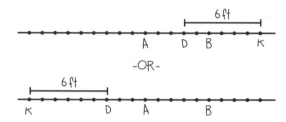

Ben and Deepa are 2 feet apart in the top drawing, and 8 feet apart in the bottom drawing.

So, the smallest number of feet Ben could be from Deepa is **2** feet.

You may also have started by drawing Amy to the right of Ben, flipping all of the drawings over to get the same answer.

70. We start by drawing an R for each of the five red blocks:

R   R   R   R   R

We draw a B for each blue block between every pair of red blocks:

R B R B R B R B R

Finally, we draw a G for each green block between every neighboring blue and red block:

R G B G R G B G R G B G R G B G R

All together, there are **8** green blocks in Kiana's row.

71. We start by drawing 10 circles to stand for our cookies.

6 cookies have chocolate chips. We add C's to 6 cookies to stand for chocolate chips.

7 cookies have pecans. We draw P's on cookies that

have pecans. We want the smallest number of cookies that have chocolate chips *and* pecans.

So, we start by putting pecans on all 4 of the cookies that don't have chocolate chips. Then, we still need 3 more cookie with pecans. These must be cookies that already have chocolate chips.

So, **3** is the smallest number of cookies that could have chocolate chips and pecans.

72. We try drawing 4 straight lines to see how many crossings we can make. The drawings below have 3, 4, and 5 crossings:

We can get even more! If every road crosses every other road, we can get 6 crossings as shown in the examples below.

Since each road crosses every other road, we cannot create more crossings than this. So, **6** is the largest number of stoplights that can be in Bantam.

73. The ☐ is to the right of the ◇ but to the left of the ◯. So, from left to right, we have ◇, then ☐, then ◯.

Since the △ is between the ☐ and the ◯, we fill the four blanks as shown below.

74. Kim is taller than Tim, but shorter than Jim. We write the three names with Jim above Kim, and Kim above Tim to show that Jim is tallest among the three.

Jim

Kim

Tim

Finn is taller than Tim, but shorter than Kim.

Finn →
Jim
Kim
Tim

So, we list all four from shortest to tallest as shown.

Tim   Finn   Kim   Jim

75. The red tub is smaller than the blue tub. So, arranging from smallest to largest, we can draw the picture below.

The yellow tub is bigger than the green tub. So, we can also draw the following picture.

Since the green tub is bigger than the blue tub, we can arrange all four tubs as shown below.

So, the **red** tub is the smallest.

76. Since T comes after P but before S, these three letters come in the order shown below.

P   T   S

Since O is not first or last, it either comes between the P and the T, or between the T and the S.

Only POTS makes a word (PTOS is not a word).

So, Tim wrote **POTS**.

77. Flynn was 3rd and Wynn was 7th. So, we label the first 7 monsters as shown:

		F				W
1	2	3	4	5	6	7

Flynn beat the monsters who placed 4th, 5th, 6th, and 7th. Wynn did not beat any of these monsters. So, Flynn beat 4 more monsters than Wynn.

We can add monsters who finished behind Wynn until the number of monsters Flynn beat is twice the number that Wynn beat.

If Wynn beat 1 monster, Flynn beat 5 monsters.

If Wynn beat 2 monsters, Flynn beat 6 monsters.

If Wynn beat 3 monsters, Flynn beat 7 monsters.

If Wynn beat 4 monsters, Flynn beat 8 monsters. ✓

		F				W				
1	2	3	4	5	6	7	8	9	10	11

The 4 monsters Wynn beat placed 8th, 9th, 10th, and 11th. So, there were **11** monsters in the race.

78. We draw lines that connect the boxes whose sum ends in 2 or 5 as shown.

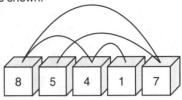

1 can only be next to 4, which must be next to 8, then 7, and finally 5. So, these boxes could be arranged in either order below.

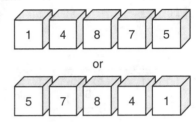

79. We draw lines that connect the boxes whose sum ends in 2 or 9 as shown.

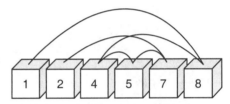

1 can only be next to 8, which must be next to 4, then 5, then 7, and finally 2. So, these boxes could be arranged in either order below.

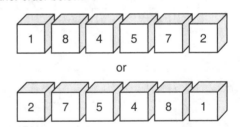

80. We draw lines that connect the boxes whose sum ends in 1 or 8 as shown.

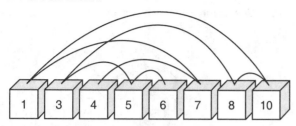

4 can only be next to 7, which must be next to 1, then 10, then 8, then 3, then 5, and finally 6. So, these boxes could be arranged in either order below.

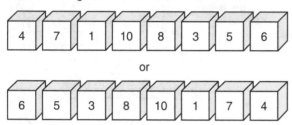

81. We draw lines that connect the boxes whose sum ends in 3 or 6 as shown.

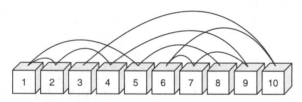

3 can only be next to 10, which must be next to 6, then 7, then 9, then 4, then 2, then 1, then 5, and finally 8. So, these boxes could be arranged in either order below.

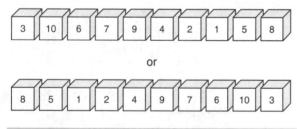

82. We draw 4 dots to stand for the 4 monsters, and connect each dot to every other dot to stand for the handshakes.

The 6 lines stand for **6** handshakes.

83. We draw 5 dots to stand for the 5 monsters, and connect each dot to every other dot to stand for the handshakes.

The 10 lines stand for **10** handshakes.

84. We draw 5 dots to stand for the 5 monsters. We label three dots H for Hatfield and two dots M for McCoy, with Matty on one of the McCoy dots.

We connect each dot to every other dot, except we do not connect Matty to any of the dots labeled H.

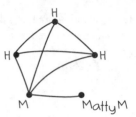

The 7 lines stand for **7** handshakes.

— *or* —

In the previous problem, we learned that 10 different handshakes are possible among 5 monsters. Since Matty refuses to shake hands with the 3 Hatfields, this removes 3 of the 10 handshakes, leaving 10−3 = **7** handshakes.

85. We draw 4 dots and label them F for Fred, G for Gary, H for Holden, and I for Iggy.

Since Fred shakes 3 monsters' hands, we connect Fred to each of the other three monsters.

Holden shakes 1 monster's hand. We have already drawn this handshake between Holden and Fred in the drawing above.

Gary shakes 2 monsters' hands. We have already drawn 1 handshake between Gary and Fred in the drawing above. Since we have already drawn all of Fred's and Holden's handshakes, the only other monster Gary can shake hands with is Iggy. So, we connect Gary to Iggy.

We now have the correct number of handshakes for each of Fred, Gary, and Holden. Adding or removing a handshake will cause one of them to have the wrong number of handshakes.

So, we use our drawing to see that Iggy shakes hands with **2** monsters (Frank and Gary).

86. We start with a medium white circle.

We can play any shape that is medium and white:

We can play any medium circle:

We can play any white circle:

So, we circle the four shapes below.

87. We start with a large gray square.

We can play any shape that is large and gray:

We can play any large square:

We can play any gray square:

So, we circle the five shapes below.

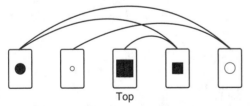

88. We label the top card, then draw lines to connect the cards that share two features.

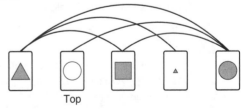

The large black square can only be played on the medium black square, then the medium black circle, then the medium white circle, then the small white circle.

So, the small white circle is on the bottom of the stack.

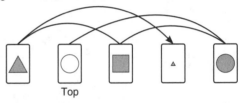

89. We label the top card, then draw lines to connect the cards that share two features.

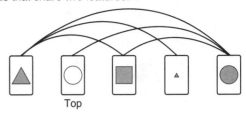

The large white circle can only be played on the large gray circle.

Then, the large gray circle could have been played on either of the other large gray shapes.

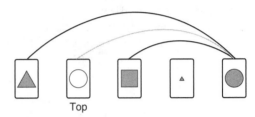

If we connect the large gray circle to the large gray triangle, there is no way to complete the stack of cards.

So, the large gray circle must be played on the large gray square, then the large gray triangle, then the small gray triangle.

So, the small gray triangle is on the bottom of the stack.

— *or* —

We draw lines to connect the cards that share two features.

The small gray triangle shares two features with only one other card (the large gray triangle). So, the small gray triangle cannot be in the middle of the stack.

Since the small gray triangle is not the top card, it must be the bottom card.

90. We label the top card, then draw lines to connect the cards that share two features.

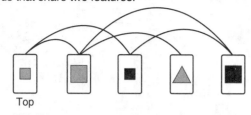

The large gray triangle shares two features with only one card. So, it cannot be in the middle of the stack.

Since the large gray triangle is not the top card, it must be the bottom card.

We check that this works by finding a path that includes all five cards, starting with the medium gray square and ending with the large gray triangle:

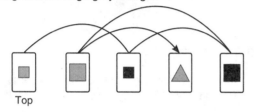

The medium gray square is on the medium black square, then the large black square, then the large gray square, then the large gray triangle. So, we circle the large gray triangle.

91. We label the top card, then draw lines to connect the cards that share two features.

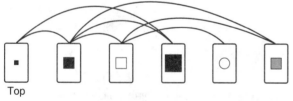

The medium white circle shares two features with only one card. So, it cannot be in the middle of the stack.

Since the medium white circle is not the top card, it must be the bottom card.

We check that this works by finding a path that includes all six cards, starting with the small black square and ending with the medium white circle.

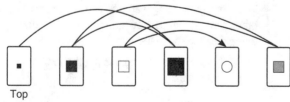

The small black square is on the large black square, then the medium black square, then the medium gray square, then the medium white square, then the medium white circle.

So, we circle the medium white circle.

PROBLEM SOLVING
Extra Information 102-103

92. Mike removed 7 cookies, and there are now 20 cookies in the jar. So, there were 20+7 = **27** cookies in the jar before Mike took some out.

We can ignore the information about time and the number of cookies Mike ate.

93. We are asked about Jane's age.

3 years ago, Jane turned 6. So, Jane is currently 9. In **7** more years Jane will turn 16.

We can ignore the information about Jane's height and weight.

94. 15 pairs of shoes is 15+15 = **30** shoes.

We can ignore the information about what kinds of shoes are in the box.

95. Since there are 3 rows, and each row has 1 more boy than it has girls, there are **3** more boys than girls in the class.

You might have found the total number of boys and girls in the class, or even the total number of boys and girls in each row, but we can solve the problem without this extra information.

96. Nickels are worth 5 cents and dimes are worth 10 cents. So, with nickels and dimes, we can only have a number of cents that ends in 0 or 5.

If the value of the nickels and dimes is 30 cents, then we must have 3 cents in pennies to make 33 cents.

If the value of the nickels and dimes is 25 cents or less, we must have at least 8 cents in pennies.

Since Grogg only flipped 7 pennies, the only way he could get 33 cents worth of coins to land heads is with **3** pennies.

We can ignore the information about the number of nickels and dimes.

97. If students got 14, 15, 16, and 17 answers *correct* on a 20-question test, the same students got 6, 5, 4, and 3 answers *wrong*.

Jeremy got twice as many wrong as Melissa. The only number of wrong answers that is double another number of wrong answers is 6. So, Jeremy got 6 wrong (14 correct) and Melissa got 3 wrong (17 correct).

So, Melissa got **17** correct answers.

We could have used the extra information to figure out that Melvin got 16 correct, and Yasmin got 15 correct, but this is not necessary for answering the question.

98. There are many ways to use 3 straight cuts to make 4, 5, or 6 pieces.

To make 7 slices, each cut must cross both of the other cuts, as shown below.

We do not recommend cutting pizza this way.

99. We do not need to find the numbers to solve this problem. Consecutive numbers come one after another, so they always have a difference of **1**.

But, in case you are curious: 193 and 194 are the consecutive numbers with a sum of 387.

100. We work backwards.

Nick is at the back of the line.

Either Lynn or Jon is in front of Nick.

If Jon is ahead of Nick, there is no one whose name ends in J to stand in front of Jon. So, Lynn is in front of Nick.

Lynn ← Nick

Either Al or Lionel is in front of Lynn.

If Al is in front of Lynn, then
Nora is in front of Al, and
Jon is in front of Nora.
This leaves Lionel out of the line.

Jon ← Nora ← Al ← Lynn ← Nick
          Lionel?

But, if we put Lionel in front of Lynn, we can put Al in front of Lionel and complete the line as shown.

Jon ← Nora ← Al ← Lionel ← Lynn ← Nick

So, from front to back, we have Jon, Nora, Al, Lionel, Lynn, and Nick.

Jon is in the front of the line.

— *or* —

There is no one whose name ends in J to stand in front of Jon. So, **Jon** must be at the front of the line.

Using this reasoning, we can ignore the information about who is at the back of the line.

101. Since 11 is the smallest number, the two remaining numbers have a sum of $37-11=26$.

So, we are looking for 2 numbers that are larger than 11 and have a sum of 26.

$12+13=25$. ✘ Too small.

$12+14=26$. ✓ Got it!

Since the numbers are all different, we can't use $11+13+13=37$.

So, the largest number is **14**.

102. Each glide takes Shelly 3 feet up the wall. Each rest takes Shelly 2 feet lower.

If we draw Shelly's path, we see that after the 3rd glide, Shelly ends up 3 feet high on the wall.

Then, on her 4th glide, she reaches the top of the wall!

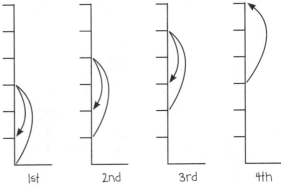

So, it takes Shelly **4** glides to reach the top of the 6-foot wall.

103. Since we can add numbers in any order, all of the expressions have the same value. So, Lizzie's expression equals the original expression:

$$(4+6)+(7+8)=\textbf{25}.$$

104. We draw 3 dots and label them L for Llama, M for Mama, and N for Nama.

There are 5 bridges from Llama Island to the other two islands. Since Mama Island has more bridges than Nama Island, we guess that more of the bridges from Llama Island go to Mama Island than to Nama Island.

If 4 of the 5 bridges from Llama Island go to Mama Island, then 1 goes to Nama Island, as shown.

This gives Mama Island 4 bridges. So, we cannot draw any more bridges to Mama Island or to Llama Island. Nama Island still needs 2 more bridges, so this doesn't work.

We try drawing 3 of the 5 bridges from Llama Island to Mama Island, and 2 to Nama Island, as shown.

Both Mama Island and Nama Island still need 1 more bridge. So, we connect Mama and Nama Island by a bridge as shown.

Now, each island has the correct number of bridges.
So, **1** bridge connects Mama Island and Nama Island.

105. We work backwards, subtracting a pack of 8 cards from Jorm's collection until he has fewer than 10 cards.

Subtracting 8 from 61 gives 53.
Subtracting 8 from 53 gives 45.
Subtracting 8 from 45 gives 37.
Subtracting 8 from 37 gives 29.
Subtracting 8 from 29 gives 21.
Subtracting 8 from 21 gives 13.
Subtracting 8 from 13 gives 5.

So, before Jorm went to the store, he had **5** cards.

106. We start with a number line, with marks drawn on it for years. We label Carly (C) and Betsy (B) 1 year from each other on the line.

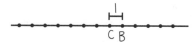

Next, we place Alicia on the line. Since Alicia is 3 years from Carla and 4 years from Betsy, we place Alicia on the line as shown.

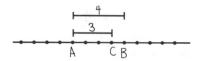

Since Denise is 6 years from Betsy and 5 years from Carla, we place Denise as shown on the line.

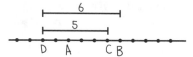

Then, since Alicia is older than Betsy, the left side of the number line is older, and the right side is younger.

We could also have flipped our number line as shown below, with older on the right and younger on the left.

So, **Denise** is the oldest.

— *or* —

We are given the age difference between every pair of children. Since Betsy and Denise are farthest apart in age, one of them is the oldest (and the other is the youngest).

We are told that Alicia is older than Betsy. So, Betsy can't be the oldest. This means **Denise** must be the oldest (and Betsy is the youngest).

107. We draw 5 dots to stand for the 5 teams, and connect each dot to every other dot twice to stand for the games played.

The 20 lines stand for **20** games.

— *or* —

We can simplify our drawing by letting each line stand for 2 games.

So, the 10 lines stand for 10 + 10 = **20** games.

108. We work backwards.

The large elefinch ate 63 peanuts.

63 is 1 more than twice as many peanuts as the medium elefinch ate. So, twice the number of peanuts the medium elefinch ate is 63 − 1 = 62 peanuts.

Since 31 + 31 = 62, the medium elefinch ate 31 peanuts.

31 is 1 more than twice as many peanuts as the small elefinch ate. So, twice the number of peanuts the small elefinch ate is 31 − 1 = 30 peanuts.

Since 15 + 15 = 30, the small elefinch ate 15 peanuts.

Together, the elefinches ate 63 + 31 + 15 = 109 peanuts. So, there were **109** peanuts in the original pile.

109. We guess the price of a pencil, then find the price of an eraser using the fact that

2 pencils + 1 eraser = 36 cents.

We check to see if our guesses work for

1 pencil + 2 erasers = 30 cents.

1st Guess: 1 pencil costs 10 cents. To make 10 + 10 + eraser = 36 cents, 1 eraser would cost 16 cents.

At this price, one pencil and two erasers would cost 10 + 16 + 16 = 42 cents. ✘ Too much!

2nd Guess: 1 pencil costs 15 cents. To make 15 + 15 + eraser = 36 cents, 1 eraser would cost 6 cents.

At this price, one pencil and two erasers would cost 15 + 6 + 6 = 27 cents. ✘ Closer than the first guess.

3rd Guess: 1 pencil costs 14 cents. To make 14 + 14 + eraser = 36 cents, 1 eraser would cost 8 cents.

At this price, one pencil and two erasers would cost 14 + 8 + 8 = 30 cents. ✓ This works!

So, 1 pencil and 1 eraser cost 14 + 8 = **22** cents.

— *or* —

We know that pencil + pencil + eraser = 36 cents, and pencil + eraser + eraser = 30 cents. So, adding (pencil + pencil + eraser) + (pencil + eraser + eraser) gives 36+30 = 66 cents.

All together, this is 3 pencils and 3 erasers. So, we can buy 3 pencils and 3 erasers for 66 cents.

To find the price of 1 pencil and 1 eraser, we need the number we can add 3 of to get 66 cents.

Since 22+22+22 = 66, one pencil and one eraser cost **22** cents.

Note that we never needed to find the costs of pencils and erasers, since we were only asked for their sum.

For additional books, printables, and more, visit

BeastAcademy.com